MW01593134

THE GOSPEL
OF THE COSMOS

THE GOSPEL OF THE COSMOS

Good News for Mankind

The heavens declare the glory of God;
and the firmament sheweth his handwork.
— Psalm 19:1

Joseph Renick

XULON PRESS

Xulon Press
2301 Lucien Way #415
Maitland, FL 32751
407.339.4217
www.xulonpress.com

Unless otherwise indicated, Scripture quotations taken from the King James Version (KJV) – *public domain.*

Printed in the United States of America.

ISBN-13: 978-1-54564-433-1

CONTENTS

PREFACE

I n the summer of 1996 a colleague of mine, Don Locker, a physicist, walked into my office one day with a copy of the June 1, 1996, issue of *Commentary Magazine*[1] opened to an article by mathematician and philosopher David Berlinski entitled "The Deniable Darwin." He pitched it onto my desk, said, "You've got to read this," and then turned around and walked out. I read it that evening at home. In this article Berlinski gives Darwinists a long, unmerciful, and thorough flogging; he does it masterfully and with style and seems to thoroughly enjoy it. So did I. Berlinski also included a positive description of the emerging (at that time) intelligent design movement that (at that time) was all new to me. In a following issue of *Commentary*[2] (Sept. 1), Berlinski's critics (mainly those whom he had flogged) responded, and then Berlinski responded to their critique. It was a thoroughly rousing give-and-take of intellectual giants slugging it out, but in the end, the effect of the second issue was to reinforce the message of the first: Darwin was *deniable*.

I had come to the conclusion many years before that chance natural processes alone could not account for the emergence of life, and Darwinism could not account for mankind. While at that time I knew little about biology and could not pose a biochemical or probabilistic argument against evolution, my conclusion came from the realization that mind and consciousness were simply beyond the reach of chance material processes. At the time that was good enough for me. Still is.

When I read Berlinski, I was at a place in my life where I was wondering what I was going to do when I grew up (I was only 58), but that evening after reading his article, I knew. Berlinski was a call to arms in a battle in which I would be engaged till my dying breath, and a major part of that battle would be to have public education quit telling children lies about the origins of mankind. It has been an exciting battle, and it is far from over, but the outcome is certain.

As I engaged in my research into intelligent design, certain things became apparent. First, it was clear that with the entry of *design* into the creation-evolution fray, the traditional conflict over the teaching of evolution and creation in public schools had suddenly become much more interesting because it introduced a different and more fundamental facet of science into the discussion that challenged Darwinism but was completely consistent with creation. Design in biology dealt with directly observed mechanisms of the life process itself embedded deep in the heart of the biological cell as opposed to speculation about unobserved past events. In the creation-evolution debate, this was a game-changer. A vector had been set in the biological sciences and it did not point toward Darwin.

Creation and *design* are natural allies but *Darwin* and *design* are natural enemies. *Creation* and *evolution* have to do with what happened in the distant unobservable past where there are inherent limitations on what can be scientifically known because there are inherent limitations on what can be scientifically observed. *Design* in biology, however, is fully observable and subject to the methods of scientific investigation; it is inferred from direct observations and we see it everywhere, every day, 24/7. In fact, we *experience* it in the very depths of our being because we ourselves are designers. We know it when we see it. We intuitively understand the concept.

It must be admitted that we can also see the effects of chance and natural selection at work in biology, but we just don't see the kinds of biological innovation (macroevolution) taking place that might get Darwin and Darwinists all excited, not because it takes place too slowly to be detected, but because chance, even acting in concert with natural selection, does not do well as a creative force in accounting for the extreme specified complexity observed in biology.

The interesting thing about the design hypothesis and the challenge it brings to Darwinism is that design is not just another version of creationism (though this is often the claim made by Darwinists), and neither was it formulated by creationists (another claim made by Darwinists). Design does not depend on the age of the earth, young or old, or what scripture may say or how scripture is interpreted. While the design hypothesis points to the work of a divine Creator, it does not depend on belief in such a Creator. It is based on directly observed evidence in biology and logical inferences from that evidence. The evidence stands on its own. Scientifically, the design hypothesis is self-evident and as a scientific hypothesis, it is *irrefutable*.

The nature of the scientific debate between Darwin and design is not over the *appearance* of design in nature. Rather it is over whether the design that is *apparent* in living things is a result of *mind* or of unguided physical laws and chance alone, where it is supposed by Darwinists that the creative power lies in natural selection. Within the biological sciences, the battle essentially reduces to two positions (pay attention here) regarding the ability of chance to account for the history of life and the origin of man: either it does because it *has to* or it doesn't because *it can't*. This is where the line is drawn. The "has to" arises from metaphysical[3] necessity. The "can't" arises from the practical limitations on what natural, unguided chance processes can do on their own.

"Has to" is a dream and "can" is a myth because "can't" is reality. Darwinists hang onto and perpetuate the myth because that is about all they have in defending their naturalistic view of the world.

Scientifically, you would think that this would be the end of the argument, but this is, at its foundation, more than a scientific argument. It is a scientific argument with profound metaphysical stakes. Further, there is more at stake than just metaphysics. Surprisingly, there are political interests at stake as well. In particular, the American Republic and its founding in natural law precepts presupposes a creator. Darwinism presupposes that there is no such creator, or, in its more diplomatic form, claims that when it comes to life, there is no *need* for a creator. This adds an entirely new dimension to the debate within public education between both Darwin and design. Not only are students being indoctrinated into a naturalistic metaphysical view of the world through the dogmatic teaching of Darwinism, they are unknowingly being indoctrinated into a secular progressive political philosophy. Of course, this is the very best kind of indoctrination, the kind that you have no idea is taking place.

Consequently, just as the implications of subject matter that touch on religious views (without specifically addressing those views) should be disclosed in public education (even though they are not), so also should the implications of subject matter that touch on political views (without specifically addressing those views) also be disclosed in public education (even though they are not). When it comes to education in biology in our public schools, there is far more going on in the classroom than science education.

❧ ❧ ❧

The intelligent design movement emerged out of a scientific dissent within Darwinism that took place in mainstream science in the 1960s, '70s, and '80s. This movement was based on discoveries in biology that came out of mainstream science itself. Since that time, a substantial volume of academic work has been produced by scientists, philosophers, and other academics, which has secured the place of the *design hypothesis* within the natural sciences, making a powerful case for *mind*, not chance, as the ultimate cause of our existence. This argument is rejected by the mainstream scientific establishment because evidence of a mind behind nature points to a divine Creator and for methodological reasons (so they say) must be rejected. Thus, unguided material causes are the only viable *scientific* alternative to giving a scientific account of our existence.

> *A major objective of this present work is to show that the design hypothesis is not only scientifically sound, its place in the natural sciences is necessary in the same way that the laws of physics are necessary. Nothing about the natural world makes sense without it. Nothing!*

Darwin and *design* support two opposing views of the world and, as will be shown, *two opposing political philosophies*. Given that only Darwinism is taught in public schools, many parents who raise their children in the Christian faith and are committed to conservative republican principles of government, have good reason to be concerned that their children are being indoctrinated into both a materialistic system of belief and a progressive political philosophy, all in the name of science. *Design* is perfectly harmonious with the natural law precepts of the Declaration of Independence and republican government, while *Darwinism* is perfectly harmonious with governments that naturally tend to be more tyrannical in nature as was so clearly demonstrated in the twentieth century.

We are left with a simple but profound question: *should the American Republic be ruled by the Laws of God or the Laws of Man?* This is what the political debate reduces to. The answer is found in the answer to another question. *Which is true, Darwin or design?* The Founders chose the *Laws of God*, and many have fought and died to preserve that principle. Today, the battleground is not the islands of the South Pacific or the beaches of Normandy. It is in public education.

The foundations of Christian faith and the foundations of the American Republic are under attack through the exclusive teaching in public schools of a naturalistic theory of the origins of man that is little more than a naturalistic fable subjected to the considerable exploratory and explanatory powers of modern science. Under the influence of special interest groups, these extraordinary efforts of government transform a glorified fable into an institutionalized lie. A fable in the hands of science is still a fable and a lie in the hands of public education is still a lie. There is simply no justification for perpetuating this indoctrination of our children. The means and mechanics for putting an end to this indoctrination are at hand. They are written into our constitution. But dear reader, no one can energize those means and mechanics but you.

∾   ∾   ∾

It was not long after I began my research before I encountered a concept within the natural sciences called *methodological naturalism*, which is a methodological principle that forbids any reference to the supernatural as a causal agent in scientific work. In science, only unguided natural causes alone can be considered in explaining natural phenomena.

As for "methodology" in science, we can think of that as having to do with *procedure, practice,* and *principle* and, in particular, the *thinking* that guides them. In several cases in the last two decades, methodological naturalism has been rolled out in defense of Darwinism, usually within the context of a dispute in public education, declaring that *design* is religion, not science and therefore unconstitutional and off limits to science education. Gaining a true understanding of the role of methodological naturalism in the natural sciences is a major objective of this work, and that will be explained in due time. Once it is exposed and understood, it becomes harmless.

Much has been written about methodological naturalism in the last three decades. Darwinists say its purpose is strictly methodological and has no metaphysical purpose. Many are skeptical about that claim and are convinced that it is applied more as a *metaphysical* constraint to protect Darwinism than a *methodological* constraint necessary to the scientific method. So far this dispute has resulted in a standoff where neither side has been able to deliver the "kill shot" against its opponent. Nevertheless, they cannot both be right. From my perspective methodological naturalism on its face seemed perfectly legitimate, and several years would pass before I finally came to understand where the deception lay.

I began writing around ten years ago, expecting to produce a modest essay on methodological naturalism of eight to ten pages, which I would circulate to others within the intelligent design (ID) community and academics who had written major works hoping that they might read it. I soon found that methodological naturalism could not be properly understood within the limited context of *Darwin versus design* but must be evaluated on more basic considerations. I also found that there was far more at stake here than an understanding of methodological naturalism. My almost fanatical sensitivity to the importance of *context* in trying to understand and resolve complex matters eventually transformed what started out to be a modest essay into this book-long treatise.

There were some big questions that needed to be answered, and those answers were important. I had picked up the scent. The hunting instinct took over. This thing took on a life of its own. I went to bed at night. It was there to put me to sleep. I got up in the morning. It was there, rearing to go. There was no escape.

I slowly began my descent into a labyrinth of issues and concerns which I would later refer to as, the "Matrix" (not to be confused with the movie). I found that everything was interconnected. An understanding of methodological naturalism required an understanding of what is meant by *natural causes*. An understanding of what is meant by *natural causes* requires an understanding of the nature of the natural sciences, and that understanding can only be found in an understanding of the philosophical and historical roots of modern science. Such an understanding in turn reveals that the logical foundation of the natural sciences rests on rational Christian theology. So what is that theology, and why is it important to science?

Further, the subject of *origins science* is inherently religious in nature, and since it is included in science curriculum and textbooks in public schools, then all such instruction must be subjected to the scrutiny of the First Amendment, which of necessity brings the judiciary into the picture. Then, if we are to understand the opinions of the courts and the judicial process, we must also gain an understanding of the judicial *philosophy* of the courts in dealing with First Amendment issues, which, in turn, is found in the accumulated history of Supreme Court decisions from early nineteenth century to the present.

Further, the First Amendment and its meaning can only be properly understood within the context of the Declaration of Independence and the founding precepts of the American Republic, such precepts being rooted in natural law. Then, given the various philosophical movements that emerged in Europe during the eighteenth and nineteenth centuries and their influence in corrupting both American politics and judicial principles, it was necessary to understand those movements and what motivated them and how they influenced both government and law in America. Then, there was the fact that public discourse concerning the evolution-creation-design controversy is conducted under the influence of a mythology, the mythology that science and religion are at war with each other. Consequently, if there was any hope for taking these ideas into the public domain, it would be necessary to dispel the warfare mythology. Then, any comprehensive treatment of evolution must include both an understanding of the evidentiary basis for the theory and an assessment of the effect it has had on religion, ethics, moral philosophy, and the progress of history in the twentieth century. Finally, all of this must be weighed in light of the extraordinary scientific discoveries in cosmology and biology in the twentieth century and what those discoveries mean for mankind.

These are the things that make up the Matrix, and this is where the answers to some of the most important questions of our times must be found, answers that affect our understanding through natural law of the relationship between men and government as well as an understanding through the natural sciences of our place in the cosmos. If we get these two things right, then, we have a "big picture" understanding our lives. We know who we are, we know where we came from, we know our lives have meaning and purpose, we know how to live our lives, we have hope for the future, and we understand the form of government that is most suited for created beings such as ourselves.

Over the years, the mental transitions back and forth into the Matrix were intense. At times, after having been out of the Matrix for several months, I would often be a bit apprehensive about going back into the Matrix because of the long intense mental effort required in the "descent" as well as the experience of sometimes getting "stuck" down there for several days. But the Matrix was territory with which I was very familiar having discovered that territory as a consequence of my involvement over the years in various high-intensity projects with the Department of Defense where we had to have answers, they had to be the right answers, and failure was not an option. The stakes were simply too high.

The same has been true here. I had to have answers, they had to be the right answers, and failure was not an option. And as you will see, the stakes are well beyond *high*.

∽ ∽ ∽

This work is described in fifteen chapters written such that the topics addressed in each of the first fourteen chapters to some extent stand-alone but also contribute to the overall narrative. Because of the maze of interconnections and relationships between topics, it was not clear just exactly how they should be organized, but I think what I have done here is reasonable. At least most of the big questions are introduced in the first chapter and the answers, such as they are, are provided along the way and then summarized in the last chapter. Hopefully, there is a reasonable flow you can follow. Regardless, it all comes together at the end.

This was where I spent much of my time during the last ten years, and the purpose of this treatise is to take you down into this Matrix and show you around. There are some troubling things down there, but there are also some amazing and wonderful things. You need to see both.

I should mention that I have a particular mindset concerning the world. I am a *Realist* and embrace what is best described as Scottish Common Sense Realism,[4] the same view that most of our Founders embraced. You will recognize what that entails as we go along. I hope you find it informative, no matter what your view of reality.

I would like to express my deep appreciation to the reader for taking an interest in these matters that are so important to our fundamental understanding of who we are, what our lives mean and how we should live our lives.

⋞ ⋞ ⋞

Given the many religious aspects of this work, it is appropriate that I disclose my own religious prejudices. I am a Christian. I refer you to C. S. Lewis' classic works[5] on Christianity for the details. During much of my adult life, I taught the Bible in my church that would be described as conservative and evangelical even though I personally tend toward the apologetic rather than the evangelical. I have always had an interest in Christian apologetics and worldview. Also since my teenage years, I have always had a sense that there was a Truth that could be known and that if you look for it you will find it, and when you find it, you will know it. I am satisfied that this same philosophy is in accord with the way the world works.

With respect to my academic credentials, I am an engineer, BS in aeronautical engineering from Texas A&M (1960), MS in engineering from Arizona State (1971), post-grad research at New Mexico Tech (1986–87) with over fifty years combined academic and work experience in science and engineering. Please understand that I am not of the academic and intellectual "class," not that anyone would ever be fooled into thinking I was. I just do not want to misrepresent my standing in the academic world. Very few in that world have ever heard of me or would care what I thought even if they had. I was able to get a good education and made reasonable (B+) grades but at Texas A&M in the late 1950s, A's in aeronautical engineering were not easy to come by.

Upon graduating from Texas A&M, I received my commission as a Second Lieutenant in the US Air Force and had the extraordinary privilege of serving nine years active duty, where as an Air Force fighter pilot, I got to "danced the skies on laughter-silvered wings."[6] Much of this time was spent on air defense alert, flying the F-102A and the F-104A. These were the Cold War years during which the bomber threat from the Soviet Union was at its peak, and we lived under the threat of nuclear holocaust. We do not ever want to go back to those times.

In civilian life, I also served several years with the Civil Air Patrol, flying the T-34A in search-and-rescue missions in New Mexico. Our two T-34s had been upgraded to the Continental 285 for high-altitude mountain searches. Mountain searches following a winter storm (which is when many of our missions were flown) are not for the weak of heart or underpowered aircraft.

My work experience following my active duty time and graduate school has been primarily in research and development for the Department of Defense, generally in the area of energetic materials, high-energy processes, blast and shock physics, gas dynamics, and weapon effects. While this treatise concerns several intellectual disciplines including science, philosophy, theology, history, education, and law, my academic work and professional career is only in science and engineering. As for the rest, I have an appetite for research no matter the topic.

❧   ❧   ❧

If you were to read just one book on the general subject of the relationship between science, religion, and philosophy, I would recommend the relatively short little book authored by John C. Lennox, professor of mathematics and the philosophy of science at Oxford University, *God's Undertaker, Has Science Buried God?*[7] Lennox is a leading Christian apologist, and in this book, he provides a good review of the nature of the metaphysical issues that are at stake in this debate. If you were to read a second book, I would enthusiastically recommend David Berlinski's *The Devil's Delusion: Atheism and its Scientific Pretentions.*[8] Berlinski is a wonderful author with a sharp wit and tongue to match. He is brilliant, he is fearless, and he shows no mercy. *Intelligent design* and *Darwinism* are both major topics in this work.

For the serious-minded, Stephen Meyer's book, *Signature in the Cell,*[9] provides a "deep-dive" into the general subject of design in the biological sciences, providing a comprehensive review of the historical, philosophical, and scientific aspects of design. Another major book by Meyer, *Darwin's Doubt,*[10] provides a thorough review of the problems with the fossil record. If you want to understand Darwinian dogma, attend an American university. If you want to understand Darwinian science, then for starters, read Meyer's two books. There are many others. The Discovery Institute in Seattle, Washington, formed in 1991, serves as the intellectual center for intelligent design—and much more. Illustra Media in La Mirada, California produces spectacular video documentaries promoting intelligent design. Excellent educational resources are available from both Discovery Institute and Illustra Media even though most are probably banned by public schools. This does not speak well of our public schools.

John Calvert, probably the foremost attorney in America regarding constitutional law and K-12 education, introduced me to the intricacies of the relationships between public education and origins science education within the context of First Amendment law. I have been much inspired by his legal analyses, his energy, and his relentless crusade in trying to get both the judiciary and public education to understand that the key to satisfying the requirements of the Religion Clause of the First Amendment in teaching origins science is simply to teach it *objectively*. And that has been the foundational principle that has guided my work ever since I first met John at a conference in Kansas City in 2001. Further, John's exhaustive legal exposition[11] on the error of the courts in their failure to adhere to a well-established judicial precedent of using an *inclusive* meaning of religion that includes atheistic and nontheistic as well as theistic systems of belief in adjudicating First Amendment cases, shows that if this inclusive meaning were applied to the relevant court cases in the last sixty years concerning the teaching of evolution in public education, all of those cases would have to be vacated and reheard. These two concepts singled out by Calvert, *objectivity in science education* and an *inclusive meaning of religion*, turns the present legal aspects of the present debate on its head. The problem today in adjudicating such cases is not with the science or its religious implications. It is with the dominant philosophy of the judiciary which gives positive law priority over natural law. As you will see, in the last 100 years, the positivist theory of knowledge has in surprising ways poisoned much of Western civilization.

I am deeply grateful to my friend and colleague, retired Sandia National Labs engineer Mike Edenburn, for long discussions over the years over hamburgers at Fuddruckers for much-needed sanity checks and the benefit of his unfailing gift for asking the right questions—no matter how irritating. I am grateful to Dr. James Campbell and the late

Dr. Ron Hadley, both retired physicists from Sandia National Labs, for their critical reviews of some of my more exploratory thoughts concerning philosophical and methodological aspects of science.

This gives you somewhat of a preview of where we are going. Stick with it. You will either really, really like the ending, or you will really, really hate it. But don't skip to the end. The journey is the best part.

≼ ≼ ≼

A note to the reader: the subject matter concerning this work is complex. Topics such as science, philosophy, and law represent distinct intellectual disciplines that have their own terminology, language, conventions, literature, and history. My own professional experience, carefully stated, is limited primarily to research in mechanics and high-energy processes, even though I have been involved in basic research in atomic physics during the last twenty years. Because I have been in science and engineering for fifty years now, research is pretty much second nature to me, so if I see something unusual floating around in somebody else's chili, the idea of poking around in their chili to see what is going on comes naturally. Much of what I am presenting here comes from someone else's chili. My personal challenge as an author is to make my arguments and present my findings in terms of concepts and language that are suitable to the discipline of concern yet comprehensible to the lay reader. For the benefit of the reader (as well as my own), the citations are not exhaustive, but hopefully they are helpful.

Design in nature is a major theme of this work—and, in fact, a crucial theme—and on that account it is fortuitous that during the past couple of years molecular biologist Dr. Doulas Axe published his book, *Undeniable*.[12] In this book, Axe (who I met briefly in Albuquerque in 2003) makes the compelling argument that we are all endowed with what he calls the "design intuition" through which we recognize things that are *designed*, and we do so reliably and without the necessity of expert knowledge. It is inherent to our basic mental operating system. Design is recognized on its face, it is self-evident, and it is *undeniable*. The design intuition is inherently reliable and is reflected in the old adage of the duck: "If it walks like a duck, quacks like a duck, and smells like a duck then it is probably a duck." Most of the time, such reasoning will get you to the right answer. Mainstream science wants us to believe that what we see in nature is only "apparent" design and not due to the work of a designing intelligence. We may hear a lot of "quacking" going and we may see a lot of "duck-like" behavior and pick up some "duck-like" smells in the world but we just have to accept that there really are no ducks out there. They are just illusions.

The strength of Axe's argument is such that we can confidently state that in the presence of intuitive design, the burden of disproof is on the skeptic. For instance, you should not, *you must not*, discount your own opinion just because it is not in accord with what your child is being taught in public-school biology classes. You have a right to express that opinion, but you would do well to do a bit of homework before you do. I humbly suggest that this treatise is a reasonable "one-stop shop" for such information. I promise you this: there are some arguments here your child's high-school biology teacher have never heard.

You will encounter Axe's "design intuition" on many pages of this treatise. Though Doug had no knowledge of my efforts, I offer him my gratitude for adding so much substance in so many places to a crucial theme of this present work.

❦   ❦   ❦

It takes a special kind of woman to live with me. I found her in 1956 and married her in 1959. My beautiful (and long-enduring and patient) wife Linda and I have been married for fifty-nine years and are still going strong. We have three beautiful, extraordinary daughters and two wonderful grandchildren. In December 2015, our youngest daughter Laura, after years and years of unspeakable physical and emotional suffering, took her own life. She was mentally ill, she knew it, and she struggled with that condition much of her life. Her departure from this life was not out of despair; rather it was with deliberation and, like a soldier who had just received deployment orders, she reported for duty. She was beautiful, really, really smart (summa cum laude in engineering), creative, and kind hearted. She was a voracious reader of the classics; loved elephants, C. S. Lewis and Narnia; outrageously hilarious; former soldier in the U.S. Army 101st Airborne Division and the fifth point in the star that is our close, loving, and somewhat crazy family. Laura was forty-three. We miss her so.

Joseph Renick

Albuquerque, NM

July 2018

CHAPTER 1

INTRODUCTION

There is a war being waged today in America. The same is true throughout the West but not like in America. Throughout the most of the West, the war is pretty much over, but not so in America where it is far from over. It is a culture war, a war of ideas between two contrasting metaphysical views of the world: naturalism and theism. These two worlviews hold eternal and absolute enmity one against the other. One view holds that we are created by God and that there is meaning and purpose to our existence. The other view holds that we are accidents of nature and that any perception of meaning or purpose we may think we see are just illusions. One view holds that there is a transcendent truth that undergirds all reality. The other holds that ideas about transcendent truth and a Creator are also illusions, just part of the Big Accident which itself is made up of lots and lots and lots of very, very, very fortuitous little accidents. We are not special. We are just lucky.

Both of these views cannot be true, but one must be true, and whichever view we might hold will have profound implications for our understanding of ourselves, our place in the universe, our obligations to our fellow man, how we live our lives, and whether we see purpose and meaning to our existence. These are not just interesting ideas that are batted around in university philosophy courses; rather, they are the kinds of ideas that cause revolutions, shape civilizations, and direct the course of history. Christianity was such an idea. So was Marxism.

The naturalistic view of the world, that is, naturalism and all its kin (materialism, atheism, humanism, secularism, etc.) relies on secular knowledge alone to inform and validate its central claims and looks to the natural sciences as the only reliable source of such knowledge. Christianity, the dominant religion in America and the West for the last 2000 years, relies principally on divine revelation for its doctrines concerning God's purposes for mankind but also claims that the natural world itself testifies (in the spirit of Psalm 19) *in a language understood by all mankind* to the fact that this is a created order.

The Christian doctrine of a created order provided the intellectual mindset that gave birth to modern science in Christian Europe in the sixteenth and seventeenth centuries. However in the nineteenth and twentieth centuries, science underwent a process of *professionalization* in which it assumed its place (and rightfully so) as a separate and independent institution within Western civilization. During this process, science also developed what might be described as an *attitude*, a mind of its own that no longer reflected the thinking of the Christian natural philosophers who laid its foundations and made the discoveries that changed the world forever. Rather, it embraced a secular view of the world that was characteristic of the cultural movements that swept across Europe during the nineteenth century and embedded themselves in American institutions in the twentieth century. Despite its roots in rational Christian theology, by

2

the late nineteenth century, those roots were being challenged, and by the late twentieth century, the natural sciences, as an institution of Western civilization, had come fully under the dominance of naturalism.

Darwin,[1] through his materialistic account of the origins of mankind, was central to the institutionalization of naturalism in the natural sciences, thereby giving naturalism a foothold in Western civilization it had not enjoyed for almost 2000 years. Darwin brought naturalists "respect"—at least in their own eyes.

This philosophical transformation resulted in a fundamental rearrangement of the relationships between science, naturalism, and Christianity. Under the new paradigm, *naturalism* has become a formidable threat to *Christianity*, *science* and *Christianity* are viewed as *natural adversaries*, and *science* and *naturalism* are viewed as *natural allies*. As in love, three is a crowd, and in *this* threesome, it is Christianity that is the outsider, the loser. Today in the West, naturalism and science are joined in unholy matrimony against Christianity.

But this is *not* the natural order of things. This is *not* the way it was in the beginning in the days of Copernicus, Galileo, Bacon, Kepler, and Newton when the philosophical foundations of the natural sciences— foundations deeply rooted in the Christian view of the world—were laid down.[2] One might wonder how you can get a naturalistic worldview out of the natural sciences that is completely contrary to the theistic worldview that provided its philosophical foundations in the first place. *How do you get naturalistic fruit from a tree with theistic roots?* You don't. Somewhere there must be a rather large disconnect in the historical and logical trail that takes you from the natural sciences, based on theistic presuppositions, to a reformulation of the natural sciences, which leads to naturalistic conclusions. This is a rational monstrosity.

This dubious heritage of naturalism is troublesome enough. But in light of the astonishing evidence of design and purpose in the natural world discovered through the natural sciences in the last 100 years, the inability of natural processes to account for life, consciousness, and mind, and the inability of neo-Darwinism to account for the origins of mankind, there is good reason to suspect that naturalism and all its kin are in existential crisis.

≈ ≈ ≈

While both Christianity and naturalism were preexistent to modern science, their individual logical and historical relationships with the natural sciences are fundamentally different. In the *science-Christianity* relationship, science was dependent on Christianity for its philosophical foundations, and once established on those foundations, it came to life and flourished. In the *science-naturalism* relationship, science and naturalism are bound together in an incestuous relationship where naturalism imposes its naturalistic presuppositions about the nature of the world on science and then looks to science to affirm those presuppositions. In some quarters, they call that circular reasoning.

Given that there is a rather large conceptual gap between naturalism as we understand it today and the natural sciences of the seventeenth and eighteenth centuries, science had to be *remade* in the image of naturalism, and that required severing science's roots in rational Christian theology. This "makeover" had little direct effect on the natural sciences in general,

4

at least regarding the mechanical workings of the universe. However, when it comes to questions that reflect on the ontological[3] nature of the world, such as the origins of the universe, life, and mankind, it was necessary to rework science's philosophical foundations to provide the right naturalistic "look." Not only was science's philosophical foundation reworked, but an oppressive oversight system called *methodological naturalism* has been put in place to ensure that evidence of design and purpose in nature is not allowed to have a voice in public school science education where young minds are shaped for the future. Material explanations, no matter how absurd those explanations may be, are mandated.

Evolutionist Richard Lewontin states this position with admirable clarity in his article in *The New York Review of Books*[4] "Billions and Billions of Demons":

> We take the side of science in spite of the patent absurdity of some of its constructs, in spite of its failure to fulfill many of its extravagant promises of health and life, and in spite of the tolerance of the scientific community for unsubstantiated just-so-stories, because we have a prior commitment, a commitment to materialism. It is not that the methods and institutions of science somehow compel us to accept a material explanation of the phenomenal world, but, on the contrary, that we are forced by our a priori adherence to material causes to create an apparatus of investigation and a set of concepts that produce material explanations, no matter how counterintuitive, no matter how mystifying to the uninitiated. Moreover, that materialism is an absolute, for *we cannot allow a Divine Foot in the door. . . .*

5

The primary problem [for science education] is not to provide the public with the knowledge of how far it is to the nearest star and what genes are made of. . . . Rather, the problem is to *get them to reject irrational and supernatural explanations of the world*, the demons that exist only in their imaginations, and to accept a social and intellectual apparatus, science, as the only begetter of truth. (my emphasis)

In spite of Lewontin's Gestapo-like advice, if we look to the *natural order* itself for an understanding of the relationships that exist between science, naturalism, and Christianity, we do in fact find a conflict, but that conflict is *not* between science and Christianity. The real conflict, as so brilliantly articulated by Notre Dame Philosopher of Science Alvin Platinga in his book W*here the Conflict Really Lies,*[5] is between naturalism and the natural sciences. In this version of the *threesome,* it is naturalism that is the outsider, the loser.

ॐ ॐ ॐ

In spite of the institutional power naturalism wields over the scientific establishment (and consequently over the natural sciences), neither naturalism nor the scientific establishment have even one iota of power over nature herself. Nature refuses to be enslaved to the naturalistic view of the world. She cares naught for the opinions of men. The discoveries by science in the last 100 years in fact reveal, not the meaningless universe of the naturalists in which mankind emerged by descent from lower forms through unguided Darwinian processes, but a universe filled with overwhelming evidence of design, meaning, and purpose. These are profound discoveries, precisely the kinds of discoveries that you would think would *excite educators* in their teaching and *delight students* in their learning about the astonishing world in which we live. You would think.

But the opposite is true. Public education presumptuously censors all evidence of design from the classroom because of its supposed religious implications. Further, it not only censors such subject matter, it also prohibits legitimate scientific critique of Darwinism, claiming that such criticism is religiously motivated and therefore is *unscientific* and cannot be allowed. As for nature, *she* could care less about religious motivation, who it is that is motivated, or what their motivation might be.

Both naturalism and public education are committed to maintaining this status quo. For naturalism, this censorship serves the crucial purpose of protecting the exclusive status naturalists now enjoy in public education and allows them to shape the curriculum of the life sciences to support the naturalistic view of the origins of mankind and thus pass on their world-view to succeeding generations. As for public education, censorship in the name of religious neutrality supports the *illusion* that public education is being neutral with respect to religion. As we shall see, nothing could be further from the truth.

∽ ∽ ∽

The Darwinian dogma is guarded by three lines of defense protecting the core doctrine of naturalism: the natural origins of mankind.

The *first line of defense* is cultural and entails a myth, the myth that science and religion are at war with each other. This myth trumps reason by casting any claims of evidence for design or criticism of neo-Darwinism as religiously motivated attacks by fundamentalist Christians on science and reason. Everyone "knows" that only a religious fanatic would actually question evolution, which is equivalent to questioning science.

The *second line of defense* is a guiding principle of science called *methodological naturalism*,[6] which assumes that science can only consider natural explanations for observed phenomena. While methodological naturalism is generally a legitimate concept for the natural sciences (to be explained later), it is misused (as will be shown) to censor evidence of design and purpose in the universe from the public school classroom.

Finally, the *last line of defense* is legal in nature. During the last sixty years in legal cases concerning the teaching of evolution in public schools, the courts have consistently ruled in favor of evolution, ostensibly in the name of religious liberty. This trend of the courts is also consistent with the general move of the judiciary toward a secular view of government and the elimination within the American Republic of any public expression that honors God. This is a very important subject which we will address in detail in chapters 13 and 14.

To put all of this in proper perspective, the reality is that Darwinian evolution as an account of the history of life and the origins of mankind, while it started out in limbo, has now descended into deep crisis. In fact, it has been in crisis (scientifically speaking) for almost sixty years, and the more we learn about biology, the worse it gets. Nevertheless, protected by the three lines of defense, public education is free to proceed unimpeded with the business of indoctrinating students into the naturalistic worldview in opposition to the prevailing theistic view of American culture and the last two thousand years of Western civilization. The central dogma of this indoctrination is the emergence of mankind from lower forms through naturalistic Darwinian processes. Evolution is mandated by state governments, and for many children (probably most), it is imposed on them against their will and the will of their parents.

So, what can be done? Not much. Given the institutional and political power held by the scientific establishment today there is little incentive for a state legislator or any other state official to make the effort to challenge this deeply entrenched status quo, which is backed by the National Academy of Sciences, national science organizations, powerful national teacher unions, Supreme Court rulings, the ACLU, local universities, liberal politicians, the liberal media, and many mainstream liberal Christian churches who blindly accept theistic evolution. Anyone who dares to take their concerns to the courts will deeply regret it. The punishment for questioning Darwin is severe.

Nevertheless, some progress is being made where (as of this writing) three state legislatures (Louisiana, Tennessee, and Florida) have successfully passed bills requiring that students be exposed to evidence both for and against evolution. In spite of this progress, we are still a long way from what might be considered as true scientific objectivity in the teaching of evolution.

There is a bright side to this battle, however. If you like a good fight against overwhelming impossible odds, this fight is for you! Let not your heart be troubled. Be of good cheer. Nature refuses to be held hostage to the minds of mere men. The naturalistic Darwinian Empire will fall. Great will be its fall, and it will be the natural sciences that take it down. The crumbling has already started.

∽ঔ ∽ঔ ∽ঔ

Whenever public education engages in instruction concerning the origins of mankind and the universe, it has touched on religion. Religion has unavoidably entered the classroom. It does not matter that the theory of concern does not make explicit reference to God as is the case for evolutionary theory. It is religious in nature because it deals with the origins of mankind. To insure religious neutrality, common sense demands that in public education when such subject matter is introduced for instruction, that students be explicitly and deliberately informed of its metaphysical or religious implications, that all related instruction be conducted objectively without metaphysical or religious bias, that such instruction be conducted with a high standard of intellectual and academic integrity, and that students be helped to understand the difference between science and a metaphysical view promoted in the name of science.

However important and reasonable you think these guidelines may be, a science teacher would probably be fired for bringing such honesty, transparency, and intellectual integrity into the classroom. In contrast to these guidelines, the approach to religious neutrality today is simply to "keep theistic religion out of the classroom," which means to ban the teaching of any aspect of the origins of mankind that might infer the work of the Divine while at the same time mandating the teaching of Darwinian naturalism. The result is the corruption of intellectual, scientific, and educational integrity and a betrayal of the trust that parents place in public education that the classroom will not be used to indoctrinate their children into any religious or political view. Political view? In science education? As we will see, the motivation to ban creation and intelligent design from

10

public school science classrooms is driven by both metaphysical *and* political considerations.

One of the principal objectives of this treatise is to expose the corruption that currently dominates origins science education and suggest a way forward in restoring intellectual and academic integrity to the teaching of the natural sciences while respecting First Amendment law. Once the bonds of naturalism (a religious view) have been broken, science teachers will be free to expose students to the extraordinary scientific discoveries that point to evidence of design and purpose in the universe. Such instruction will entail principles that are in perfect accord with the opening paragraphs of the Declaration of Independence concerning the laws of nature and of nature's God. What possible objection (other than a dislike of our founding precepts and disbelief in God) would government have to the idea that the natural sciences affirm the founding principles of the American Republic?

Much is at stake. Darwinism, as a scientific theory of the natural origins of mankind, is for all practical purposes dead. However, as a secular religion it is alive, well and thriving with a bright future where Darwinism is propped up and paraded around as if all were well (the dark 1989 comedy movie *Weekend at Bernie's* comes to mind). But there is no comedy here. The future of Darwinism and, consequently, naturalism depends completely on the ability of the scientific and educational establishments to maintain their iron grip on origins science education. Such control is easy to maintain in a totalitarian state such as the former Soviet Union but impossible in a free society, such as America. Nevertheless, freedom of religion is being eroded within our public institutions, and nowhere is that erosion as evident as in public education. Accordingly, the scope of this work extends well beyond the question of science education

11

in public schools to the broader impact of Darwinism as a secular religion and its implications for the preservation of not only religious liberty but the American Republic itself.

∾   ∾   ∾

In the last chapter of his book *Evolution: A Theory in Crisis*,[7] Michael Denton discusses how evolutionary biology has impacted Western culture in ways that go far beyond the boundaries of science.

The twentieth century would be incomprehensible without the Darwinian revolution. The social and political currents which have swept the world in the past eighty years [Denton is writing in 1986] would have been impossible without its intellectual sanction. It is ironic to recall that it was the increasingly secular outlook in the nineteenth century which initially eased the way for the acceptance of evolution, while today it is perhaps the Darwinian view of nature more than any other that is responsible for the agnostic and skeptical outlook of the twentieth century. What was once a deduction from materialism has today become its foundation.

The influence of evolutionary theory on fields far removed from biology is one of the most spectacular examples in history of how a highly speculative idea for which there is no really hard scientific evidence can come to fashion the thinking of a whole society and dominate the outlook of an age . . . One might have

12

expected that a theory of such cardinal importance, a theory that literally changed the world, would have been something more than metaphysics, something more than a myth.

Ultimately the Darwinian theory of evolution is no more nor less than the great cosmogenic myth of the twentieth century . . .

The truth is that despite the prestige of evolutionary theory and the tremendous intellectual effort directed towards reducing living systems to the confines of Darwinian thought, nature refuses to be imprisoned.

This excerpt from Denton captures very well the historical and cultural influences of Darwinism on Western civilization since late nineteenth century.

The contribution of Darwinism to scientific knowledge has been of relatively little significance, but its naturalistic account of the origins of mankind has shaken the teleological[8] foundations of Western civilization. We are nearing the end of the sixth decade since the triumphant Darwinian centennial celebration held at the University of Chicago in 1959 and, as indicated by Denton, it is safe to say that today the naturalistic metaphysics of neo-Darwinism has in one way or another influenced just about every institution of Western civilization. To suddenly overthrow the Darwinian paradigm, in the words of Thomas Woodward in his book *Doubts about Darwin*,[9] "would send shock waves through the secularized bedrock upon which so much of our civilization is built." Undoubtedly, what Woodward

13

says is true, but it may be only then that we would fully comprehend the depth and breadth of the Darwinian malaise that pervades Western civilization.

⋙   ⋙   ⋙

Ultimately, secular knowledge based on scientific discovery will come down on one side or the other of the question: Is our existence the result of unguided material causes alone, or is it due to the work of a guiding intelligence? *The cosmos holds the answer to this question*, and ever since man first lifted up his eyes to the heavens, it has been giving us hints, speaking to us in a language understood by all nations.

Here in early twenty-first century, scientific knowledge has progressed to the point where the answer to this question, an answer that resonates with Psalm 19, is no longer in doubt. The scientific establishment will vehemently disagree with this conclusion, but their objection is based on their naturalistic commitment, not an objective consideration of the plain facts. While naturalism in public school science classrooms is, for the present, secure, its days are numbered, and science will be its downfall. This is inevitable because science is the study of nature, and as Denton points out, nature refuses to be confined to naturalistic thought. The scientific pendulum is swinging with irresistible force back to the teleological view of the world, and as it swings it will sweep away the soul-sucking naturalistic parasite that has been so deeply embedded in the natural sciences and our culture. We will finally ask "What were we thinking? How could we have been so blind? Why was the deception so deep, so broad, and so enduring?"

Of particular interest to this work is the impact of Darwinism on the American Republic where, as we shall see, there is a fundamental conflict between Darwinism and the precepts of the American founding. If there is no place for Darwinism in the Declaration of Independence, there is no place for Darwinism in the American Republic. It really is that simple. Yet today, in public education, God is out and Darwin is in.

CHAPTER 2

METHODOLOGICAL NATURALISM, CONSTITUTIONAL LAW AND SCIENCE EDUCATION IN PUBLIC SCHOOLS

M ethodological naturalism[1,2] is a guiding principle imposed on scientific methodology that assumes that the natural world is fully explainable in terms of natural causes alone without appeal to the supernatural. The claim is made that this methodological assumption is necessary because science is inherently limited by its methods to the investigation of only that which can be observed, and since science cannot observe supernatural causes, it must rule out such causes, not as a metaphysical predisposition but as a practical matter. Further, science would be impossible if some supernatural agent was continually tinkering with various properties of the natural world, and since there is no evidence of such tinkering going on, we may conclude that either the supernatural does not "tinker" or there are no supernatural causes involved in the workings of the natural world—at least to the extent that they produce effects observable through the methods of science.

John Rennie, editor in chief of *Scientific American* provides a concise statement about methodological naturalism in his article "15 Answers to Creationist Nonsense."[3] The title of his article tips us off as to his views on creation.

> A central tenet of modern science is methodological naturalism—it seeks to explain the universe purely in terms of observed or testable natural mechanisms.

The prevailing view in science is that life and mankind simply emerged from preexisting nonliving materials through natural unguided processes. According to this view, both life and mankind must be considered as being part of the natural world where there is nothing about the existence of either that is fundamentally different from anything else in the natural world, including *mind*, which also emerged from matter. This position, while part of the naturalist's dogma, is controversial, especially within the cognitive sciences. For instance, consider an excerpt from philosopher Thomas Nagel's great little book *Mind and Cosmos*[4]:

> The existence of consciousness is both one of the most familiar and one of the most astounding things about the world. No conception of the natural order that does not reveal it as something to be expected can aspire even to the outline of completeness. And if physical science, whatever it may have to say about the origin of life, leaves us necessarily in the dark about

consciousness, that shows that it cannot provide the basic form of intelligibility for this world. There must be a very different way in which things as they are make sense, and that includes the physical world, since the problem cannot be quarantined in the mind.

While Nagel draws the explanatory limit of naturalistic science as presently understood at the level of consciousness, there is a convincing case that can be made that this limit actually exists at least at the boundary between the living and the nonliving world—and maybe all the way back to the very existence of the natural world itself. There is compelling evidence that such is the case—a case that is established through observable facts without any need to invoke the supernatural. Nagel might even agree to the living/nonliving boundary as suggested by his reference to Darwin in the subtitle to his book, *Why the Materialist Neo-Darwinian Conception of Nature Is Almost Certainly False*.

Today, despite the smug confidence of naturalistic scientists and philosophers, naturalism faces an uncertain future. We shall see that in the face of a growing metaphysical crisis within science as described by Nagel (and denied by almost everyone else) methodological naturalism, while serving the legitimate role of guarding natural causation in scientific methodology, has been given a covert and impossible mission: that of defending naturalism from evidence of design and purpose in the universe.

≈ ≈ ≈

Iris Fry, a philosopher with the Department of Humanities and Arts at Technion-Israel Institute of Technology, wrote an article published in the journal *Studies in History and Philosophy of Biological and Biomedical Sciences*,[5] "Is science metaphysically neutral?" As indicated by the title, she challenges the notion that methodological naturalism is a metaphysically neutral concept. Fry, who discloses her own naturalistic view and her conviction that science supports that view, provides a brief account of the historical transformation of science from a theistic metaphysical framework to the naturalistic metaphysical framework. Darwinism was central to that process. Here are key excerpts from Fry's article. Her candor is refreshing.

It is argued that methodological naturalism, the notion that science can only pursue natural explanations for phenomena observed in the natural world, is metaphysically neutral having nothing to say about the existence or non-existence of a transcendent creative power. However, there is more to it than that. Behind the commitment to naturalistic methods is the conviction that life emerged according to unguided natural means alone and this dogma is central to the naturalistic worldview. The claim that the theory of evolution is metaphysically neutral does not stand serious philosophical and historical scrutiny. Darwinism was crucial in the historical change from the understanding of life in the framework of a theistic metaphysics to the establishment of a naturalistic worldview. By telling us that God was not necessary, Darwinism is anything but metaphysically neutral.

Asserting the metaphysical neutrality of science is considered an 'essential strategy' in order not to antagonize the
majority of Americans who are believers and to win them over
to evolution.

Fry spills the beans. Science (in her view) is anything but metaphysically neutral. *Naturalism* as a worldview depends on science for its
intellectual foundation; consequently science must be conducted solely
within a *naturalistic* framework. In a world overrun with life, design,
and purpose, this is naturalism's only hope. Naturalism provides the
metaphysical foundation of the natural sciences, and in return, the natural sciences provide a bit of life-sustaining oxygen to naturalism. In its
modern manifestation, science is indistinguishable from naturalism, and
naturalism cannot survive as a worldview if evolution does not survive as
a scientific theory of the origins of mankind.

Fry says that as an "essential strategy," science education must make
a pretense of religious neutrality all the while working to win student
"believers" over to naturalistic evolution. Don't worry about trying to
convince the Christians in their churches. Go after their children in public
schools. This was the way the communists did it in China after the revolution in 1949, and it is the way it is being done in America today.

One of the most fundamental questions we can ask is . . .

> *Is our existence the result of unguided material causes alone, or is there a guiding intelligence behind it all?*

This is an inherently religious question in that regardless of the answer we may come up with and regardless of how that answer is obtained—whether scientifically or otherwise—it will have profound implications for how we view our lives, our understanding of our place in the universe, how we should live our lives, how we should treat our fellow man, our view of eternity, and whether there is meaning and purpose to our existence. Consequently, it is unavoidable that any scientific theory that concerns itself with human origins will have profound religious implications. Thus when public education chooses to instruct students in scientific theories concerning the origins of mankind, it has engaged in an activity that "touches" on religion. *Religion has unavoidably entered the classroom*, and it is a great offense to students and their parents for public education to pretend that it hasn't.

The courts, in consideration of the First Amendment, have ruled that whenever government is engaged in an activity that "touches" on religion, it must be *neutral with respect to religion* (details later). Even if a theory of origins is not explicitly religious in nature, that is, it does not depend on some religious premise or source, it is still accompanied by religious implications, implications from which it cannot be separated. Consequently, public education is confronted with a question (which it completely ignores) that goes to the very heart of the concern so many Americans have over just exactly what it is that public education is teaching their children in biology classrooms.

21

> *How does public education maintain neutrality with respect to religion while instructing students in scientific subject matter that teaches children that they were not created by God, but are accidents of nature?*

While public education today will say *nothing* one way or the other about the existence of a divine Creator, there is a subtle but rather effective form of indoctrination adopted by public education which communicates its views regarding creation by saying *nothing*. By saying *nothing* about *something*, the message that is silently and insidiously conveyed is that such an idea is unworthy of mention, it is irrelevant, it should be ignored, it is shameful, and eventually it must fade from memory. Such is the intended fate of *Creation* and *intelligent design* in public education. While they are the "elephants" in the biology classroom that everyone sees, no one dares mention them. This is metaphysical indoctrination, not biology education.

There is little evidence that any public school system in our country fully understands the complexities of the issues that surround these questions. In fact, the same thing could probably be said about the judiciary. The reason in part is due to the way science defines "natural causes" and the way the judiciary defines "religion." We briefly address the legal meaning of religion here. Chapter 6 is dedicated to explaining natural causes.

The courts have adopted an *inclusive* meaning of religion, which includes atheistic and nontheistic systems of belief as well as traditional theistic systems of beliefs. The adoption by the courts of this *inclusive* meaning of religion is treated in considerable depth by Kansas City attorney John Calvert in his review of *Kitzmiller*.[6] By employing the *inclusive* meaning of religion, government supposedly guarantees equal treatment under the law regardless of religious belief, whether it be atheistic, nontheistic, or theistic. While there are a number of *Free Exercise Clause* cases (see Calvert) in which the inclusive meaning of religion was a major factor in adjudicating those cases, that meaning has *never* been incorporated in any of the *Establishment Clause* cases heard by the courts dealing with the teaching of creation, evolution, or intelligent design in public schools. In all of these cases, the use of an exclusively theistic meaning of religion was essential to the reasoning of the courts in reaching their conclusions that have secured, under false pretenses, the *evolution-only* doctrine within public education. The rationale for their findings in every case rested on the reductionist premise that creation and intelligent design are "religion" and evolution is "science."

But in spite of the influence of materialism, bad assumptions, wrong turns, and more bad assumptions, in time science itself will figure all of this out because that is what science does. Dogmas come and go, and sometimes they stay around far too long but, as Michael Denton said, "nature refuses to be imprisoned" by those dogmas, and eventually scientists will, through the ancient process of "changing of the guard," finally "get it." Eventually, the *old guard* will pass away and with it, the *old illusions* about the seemingly miraculous powers of chance processes. This seems to be a generational sort of thing, and in due time, corrections will be made. One of those corrections will be to exorcise the curse of Darwinism from the world. It is just a matter of time, and when that time comes, perhaps historians will be inclined to finally tally the awful cost of Darwinism to mankind.

Somehow, I would like to believe that the scientists of the *new guard* are in the world today, waiting for "their time" and that this *new guard* will reject the tyrannical dogmas of the *old guard*. Their strategy is simple. Restore intellectual and academic integrity in the natural sciences and then follow the evidence wherever it may lead. Just try it. I think you will really like it.

∽ ∽ ∽

Providing a legitimate and practical answer to the question framed above concerning the teaching of origins science in public education that respects both the religious and political commitments of students and their parents is a major objective of this treatise. The pursuit of this objective requires that we consider a wide range of relevant topics in establishing the scientific, educational, legal, philosophical, religious, and historical framework within which this answer must be found. And herein resides the problem. Policies and practices today are posed within a narrow and distorted framework of ideas built upon mythologies and misrepresentations driven by naturalistic commitments within science, progressive ideals within politics, a positivist legal philosophy in law, and a secular state. Any attempt to expose and then tear down this biased framework and replace it with an objective, unbiased, and authentic framework that reflects the facts will be met with fierce scientific and political resistance.

For too long, many Americans have walked away from this fight, surrendering cultural power to the enemy. They have either viewed it as a hopeless fight that has already been lost, or for religious reasons, they thought that they should separate themselves from the evils of the world by staying out of that fight. Unfortunately, most do not understand the nature of the battle and the spiritual and political stakes involved. No one with metaphysical, political, or cultural stakes in Darwinism is going to walk away from this fight, no matter what.

It is my hope that one result of this treatise is that many will come to understand that we are involved in a great battle, that battle is hardly over, and even though the outcome is assured, it matters tremendously whether or not you are in the fight. To do nothing is to surrender, and surrender is not an option—not for me.

৶   ৶   ৶

We now turn to the task of laying out a framework of concepts and principles for guiding educational policy and practice. Once this framework is understood, the way forward in setting things right will become self-evident. However, there is a complication. Someone has "monkeyed" with the framework itself to accommodate a certain perception—the perception that naturalism, the natural sciences, and evolution together make sense and provide the only rational way to look at the world.

Within the present framework, the corrective actions, however contemptible they may be to some, are simple and straightforward, protecting religious and political rights, restoring integrity to science and science education, regaining the trust of parents, and honoring the natural law principles of our founding fathers. But the framework, itself, will resist such an outcome.

Thus, the way forward must ultimately require correctives at two levels. The first is correctives within the present framework, the second is correction of the framework itself. Neither will be easy, but restoring the cultural framework will undoubtedly prove to be difficult. It may take another "Greatest Generation" to pull it off, and it remains to be seen if there is anything like that left in this great American experiment of which we have been a part. If not, the experiment is over. It will have failed. But not on my watch.

CHAPTER 3

THE NATURAL SCIENCES

I do not wish to engage in an overly complicated discussion about science and its methods, and it certainly is not my intention to give the impression that what follows is a final authoritative statement about what *the scientific method* is. Most scientists and philosophers of science are convinced that there is no universal definition of the scientific method. There are, however, certain presuppositions, methods, procedures, and ways of thinking that are necessary to the practice of science.[1, 2, 3, 4, 5] Whatever the role methodological naturalism might play in the natural sciences, this is where we will find out what that role actually is.

Modern science arose in Christian Europe in the sixteenth and seventeenth centuries, there and nowhere else. It was the rational Christian doctrine of a knowable created order that made this possible. No other philosophical or religious system of thought, including atheism, expressed a worldview that allowed for the unique mindset necessary for science to emerge. This does not mean that modern science literally emerged *from* Christianity itself. It just means that certain philosophical premises are required for the practice of science, and those premises were preexistent in the Christian view of the world—and no other, not then and not now.

The Christian view of a created order leads to the presupposition that there is an *objective reality* that exists independent of the opinions, preferences, and beliefs of man. That reality existed long before man arrived on this earth, it will continue to exist long after he is gone from this earth, and while he is here on this earth, there is nothing he can do to change that reality. Mankind, whether he likes it or not, is an essential part of that reality.

Further, it is necessary, and thus we presuppose that the rationality of the natural world and the rationality of the mind of man are the same rationality, and we know that they have the same rationality because they have the same Author. Consequently, the universe is *intelligible* to the mind of man, and science is possible. This is a great mystery for naturalists but a self-evident truth for Christians. Christian natural philosophers not only held that the natural world was intelligible, but it *invites exploration*. It was meant from the beginning that we come to an understanding of God's created order. Their principle motivation was the conviction that by gaining knowledge of the natural world, they were gaining knowledge of God.

Today, however, things are fundamentally different. A major priority of the scientific establishment is to use scientific discovery to *dispel* belief in God, and the focus of this priority is our public schools.

The early pioneers of science like Copernicus, Galileo, Kepler, Bacon, Descartes, Boyle, and Newton adhered to the Christian patristic[6] writings that held that there are *two books* that lead to true knowledge—the *book of God's Word*, which provides knowledge of God and reveals His plans and purposes in redemption, and the *"book" of nature* (or the "book of God's works"), which provides knowledge of the natural world and reveals God's works in creation. Both books are read through the eyes of faith and reason and because they have the same Author, they cannot be in conflict. While naturalistic scientists and philosophers would strongly disagree with this portrayal of how faith and reason work together, hand in hand, in discerning true knowledge, it is nevertheless true that it was within this perspective that modern science developed.

It is only through a *true* history of science (as opposed to a *fabricated* history) that we come to an understanding of the actual historical connection between science and Christianity. What this history teaches us is that *faith* and *reason* are inseparable. Why, in a created order, should it be otherwise? Both are gifts from God, the "eyes" by which we "read" the *two "books"* and come to a knowledge of the Creator.

Note that both the theistic and the naturalistic views of the world have stakes in the *book of nature*. *Both* look to this *book* for illumination and as a source of knowledge to support their distinctive views of the world. While both lay claim to the *book of nature* as their own, ultimately it can only support one of those views. Rightful ownership is established by what the *book* itself has to say, not what naturalistic scientists have to say. Naturalism demands that everything, including life, must be explainable in terms of natural causes alone. However, if there is anything substantive about the natural world that cannot be explained by unguided natural causes alone, then anyone who depends solely on *that* version of the *book of nature* will unavoidably be left with an incomplete view of the world. That, incidentally, was the point made by Thomas Nagel.[7]

An interesting question presents itself to those who would explore the book of nature.

> *Does the book of nature point to a cause beyond itself?*
> *Is it possible that the natural world might reveal the*
> *"fingerprints" of the Creator?*

A simple reading of the book of nature not only reveals the works of a transcendent creator, but, as will be shown, his plan and purpose in creation as well. This poses a huge dilemma for naturalistic scientists. How do naturalists deal with this dilemma? Do they deny it? Do they ignore it? Do they attempt to explain it away? Or, do they face it honestly and give serious consideration to the merits of such evidence?

Because of the overwhelming nature of the evidence of design and purpose, it cannot be ignored, and neither can it be denied. It must be dealt with, and science does so by attempting to explain it away, claiming that evidence of design and purpose in the natural world is an *illusion* that results from unguided chance processes alone. The principle naturalistic theories employed by naturalists to account for evidence of design and purpose in the natural world are *Darwinism* to give a naturalistic account of the history and diversity of life and the emergence of mankind, and *multiverse* to give a naturalistic account of a cosmos inexplicably finely tuned for life. The essential principle that underlies both is *chance*. As absurd as it may seem, in the world of naturalism, *order* comes from *chance*, the laws of nature come from chance, *life* comes from *chance*, *consciousness* comes from *chance*, and *mind* comes from *chance*. We are all the result of blind, uncaring, *chance* processes. Obviously, this is not a conclusion that can be inferred from *empirical evidence* or deduced from *scientific principle*. It is a matter of *metaphysical necessity* because to the naturalistic mind, the alternative is unthinkable. This is the mindset that rules science education in America's public schools.

∾   ∾   ∾

The discipline of science is concerned with the pursuit of knowledge about the natural world. It involves certain methods and practices such as observation, experimentation, and analysis conducted for the purpose of gaining an understanding of the nature of, principles behind, and physical laws, which govern the operation and behavior of the natural world. Information derived from these methods is generally referred to as "scientific information," which includes observation, data, evidence, logical inferences, hypotheses, theories, physical laws, regulating principles, causal mechanisms, conclusions, and any other classification of information that results from scientific work. The aim of scientific inquiry is to discover the underlying laws, principles, and mechanisms that account for observed physical phenomena, to rigorously explain that which is observed, and to get as close to the truth as possible, whatever that truth might be. Ultimately, nature's secrets can only be discovered by allowing the evidence to speak for itself.

It is observations and experiments that provide the connection between the abstract mental constructs of science such as hypotheses, theories, principles, and laws, and the actual behavior of the natural world. *Observation is what grounds scientific findings in reality*. Physicist Richard Feynman affirmed this principle when he said[8]:

31

The principle of science, the definition, almost, is the following: *The test of all knowledge is experiment.* Experiment is the *sole judge* of scientific "truth."

❧   ❧   ❧

While it is observation and experiment that ground science in reality, when it comes to the study of events and processes that took place in the remote past, a different kind of thinking is required in that these events and processes are inaccessible to either observation or experimentation. Such scientific investigations make up a unique category of science called the *historical sciences.* In such investigations, the challenge is how to develop a scientific understanding of *unobserved* events and processes of the remote past. Given the long debate in science from the seventeenth century to the present over *certainty of knowledge* (to be addressed in more detail shortly), we are presented with a question: How do you establish certainty of knowledge concerning events and processes that cannot be observed and for which experiments cannot be performed? How do you *really* know *for sure* what happened? You don't. You don't *really* know because you *can't* really know, and the integrity of science demands that if you don't really know, you should admit that you don't really know and quit making up *stuff* and then imposing it on students in public education as undisputed scientific fact.

These limitations on the historical sciences are unavoidable. However, in spite of the challenges posed, there are ways to arrive at an understanding of singular past events and processes even though such an investigation can never conform to the same rigorous norms of scientific investigation as typically practiced in physics and chemistry, that is, directly observed phenomenon and repeatable, reproducible experiments. The details of such past events are simply not accessible to this mode of investigation. Accordingly, while useful information might be obtained, it is impossible to establish certainty of knowledge equivalent to that obtained in physics and chemistry, no matter how many different *lines of evidence* (something evolutionists like to brag about) might be employed. It may be possible to gain *consensus of opinion*, but for the historical sciences, a mainstream consensus of opinion is likely to be more an indicator of institutional bias than what the evidence has to say. This bias is pretty much guaranteed if metaphysical claims are at stake. Such is the case for Darwin's theory of evolution.

Origins science, that is, the study of the origin and development of the universe, life, the diversity of life, and mankind is by its very nature *historical* where scientists attempt to reconstruct a historical narrative that gives an account of unobserved events and processes of the remote past, which took place long before man appeared on the earth. What we must remember is that while scientists may be able to construct plausible narratives about the past, because of their inability to directly observe what actually happened, it is impossible to confirm that such narratives are true, even if there are lots of them, there are lots of scientists that believe them, they have been written into millions of textbooks, and for the last sixty years taught as undisputed fact. These historical narratives, which are constructed on a seemingly endless supply of circumstantial evidence (fondly referred to as *lines of evidence*), are particularly vulnerable to naturalistic prejudices.

This unavoidable feature of evolutionary science is problematic in that the key aspects of scientific methodology, *observation and experiment*, are severely constrained. You may be able to directly observe fossils, the leftovers of life, and the only direct physical evidence concerning life's history. You may collect pertinent data pertaining to their geological and geographical sources, and you may be able to infer something about their age and general historical and anatomical relationships, but you cannot directly observe the *events, mechanisms and processes* themselves that produced the living creatures those fossils represent and which are the very things you set out to explain in the first place.

The late Ernst Mayr, in his memorable article "Darwin's Influence on Modern Thought," published in *Scientific American*,[9] summarized all of this with an honesty that is somewhat inspiring when he wrote:

> Darwin introduced historicity into science. Evolutionary biology, in contrast with physics and chemistry, is a historical science—the evolutionist attempts to explain events and processes that have already taken place. Laws and experiments are inappropriate techniques for the explication of such events and processes. Instead one constructs a historical narrative, consisting of a tentative reconstruction of the particular scenario that led to the events one is trying to explain.

Because of this inherent limitation of the historical sciences, scientists must be particularly careful to avoid making unsubstantiated claims that go far beyond the evidence—or, as often is the case, against the evidence. It is easy for *historical* science to become *sloppy* science, where without even being aware of it, the scientist imposes a metaphysical conclusion on the evidence so that it conforms to the prescribed view. Much worse, however, especially in public education, is the deliberate misrepresentation, corruption, or omission of evidence to ensure students come to the "right" conclusion. *This is indoctrination, not education.* It is also typical of the Next Generation Science Standards[10] and high school biology textbooks.[11]

Thus, in the historical sciences, scrupulous objectivity and intellectual integrity are required in producing an objective treatment of the evidence, especially when there are metaphysical commitments at stake or the evidence points in a direction that is inconsistent with the *accepted*, or *consensus* view. This objectivity and integrity are especially important when the context is public education and the subject matter has unavoidable religious implications.

In spite of the paucity of evidence supporting Darwin's two theses, *common ancestry descent* as the pattern for the history of life and *natural selection* as the dominant mechanism for biological change, and in spite of the fact that neither can be subjected to Feynman's test ("experiment is the sole judge of scientific truth"), and in spite of the fact that both fail Rennie's criteria under methodological naturalism ("observed or testable natural mechanisms"), their scientific status is secured by the consensus of a scientific establishment committed to a naturalistic metaphysics and (as you will see) a progressive political philosophy.

❧ ❧ ❧

The Englishman Francis Bacon (1561–1626) had tremendous influence on development of the methods of scientific reasoning. A devout Christian, he believed, along with other natural philosophers of his day (today we call them scientists), that God created the universe and put in place physical laws to govern its behavior. God was *Lawgiver* over all of creation—for both mankind and nature. In Bacon's view, natural philosophers should assume as a working hypothesis that all observed phenomena can be explained in terms of the Creator's natural laws. Note the similarity between methodological naturalism and Bacon's view—except, of course, the part about the laws of nature being laid down by the Creator.

The historical context of Bacon's guidance was that natural philosophy had been imprisoned throughout the Middle Ages by Aristotelian philosophy, which was regarded by the scholastics in the universities of Europe as unassailable truth. The official position within the universities was that all scientific work must revere the ancient authorities, and none were more revered than Aristotle. If you were an academic and your work was to be accepted in the academic world of those times, it had to be in accord with Aristotle.

Bacon, along with Galileo, Kepler, and Descartes emphatically disagreed. Bacon argued that we should not look at the natural world through any metaphysical or authoritative "lens," even one with the standing and respect of Aristotle. Set aside any preconceived notions. Look to the natural world *itself* for understanding. Let the natural world speak for itself.

Bacon's logic was *inductive*—going from the "particular" to the "universal" rather than *deductive*—going from the "universal" to the "particular." For almost 2000 years, Aristotle had been accepted as the "universal," and Bacon and Descartes recognized that this philosophical system was a dead end that had to be abandoned. But what was the problem with Aristotle?

Aristotle held that matter was eternal and was given certain forms and functions as determined by the *demiurge*. The result was the appearance of order, regularities, and harmony in nature that seemed to be directed toward some intended end. This was the basis for the *teleological* view held by Aristotle and the Greek stoics.

Aristotle developed a *unified* theory of metaphysics and cosmology. Ideas about moral perfection and corruption were intimately connected to the workings of the material world. According to Aristotle, matter was eternal, but it was also *corrupt*, and because of that corruption, its natural tendency was to sink downward toward the center of the cosmos. Why? Because that is what corrupted things do. The circularity in the reasoning here is unmistakable. "Heaviness" was a measure of corruption. Corruption was the *cause*, and heaviness was the *effect*. Things behaved the way they did because of inherent tendencies. It was their "nature" to behave that way. Seemingly, there were immaterial causes at work in Aristotle's world where explanations involved occult *forces* rather than natural *mechanisms*.

These ideas were impressed on the cosmos where the earth, made up of matter, was at the center of the universe. According to Aristotle, man lived in the sludge at the bottom of the drain, a cosmic cesspool. The perfected realm was the realm of the heavenly bodies, consisting of everything outside the sphere of the moon. These bodies circled the earth in perfect circular orbits because the circle was the perfect shape, and everything in the heavenly realm was perfect.

For centuries, the Aristotelian system with its embedded tautologies, occult forces, geocentric view of the world, and search for *causes* threw natural philosophers off the trail of a true understanding of the natural order. There was simply no place for the concept of natural law and mechanism within the Aristotelian system. Further, much of Aristotle was contrary to Christian doctrines, which held that the cosmos was not *eternal* but *created* and, as we are told in Genesis, when God completed His creative work, he saw that it was "very good"—not corrupt. That came later.

This then was the historical and philosophical context of Bacon's advice to natural philosophers. They should rely on observational data from the natural world to guide their investigations, not deductions from any metaphysical view or authority.

It was Copernicus and Galileo that made the first big break from Aristotle with their heliocentric view of the world. Galileo, guided by his telescopic astronomical observations and the prior work of Copernicus, was able to break away from Aristotle's geocentric dogma while retaining (mistakenly) his circular orbits. Kepler, through analysis of the extensive astronomical data of Tyco Brahe, finally established that the orbit of Mars was not circular but elliptical (actually, it was very slightly perturbed from the ideal elliptic form of two-body dynamics due to the gravitational influences of the other planets, especially Jupiter and Saturn). Thus another Aristotelian dogma crumbled. Once freed from Aristotle, science came into its own.

We note at this point that while science was set free from the Aristotelian metaphysical system in the seventeenth century, here in the twenty-first century, it has come under the rule of a new metaphysical system—*naturalism*.

⁊ ⁊ ⁊

Having mentioned Bacon and his *inductive* approach to scientific reasoning, it should be said that *induction* as Bacon applied it, was certainly *necessary*, but it was *insufficient* in and of itself for a complete system of scientific reasoning that would lead to certainty of scientific knowledge. There was a debate going on during this period over how to achieve "certainty of knowledge." It was held that there was an objective reality that could be known, and the kind of knowledge produced by science with its reliance on observation of the natural world itself (the Baconian approach) made it possible for scientists, at least in principle, to draw near to the truth about that reality. The insufficiency of Bacon's approach alone was that it would be impossible to establish universal knowledge with great certainty based on discrete, or particular, observations or experiments. The problem was basic. How do you get to the universal, the physical laws and principles of nature, from the particular, the individual observations and experiments demanded by Bacon?

Rene Descartes, a devout Catholic, French philosopher, and younger contemporary of Bacon, like Bacon, had a deep aversion to Aristotle. Descartes sought to *explain* natural phenomena in terms of mechanical or natural causes that were present within nature rather than *describe* them in terms of Aristotle's qualities and forms. His approach to knowledge was to attempt to prove that those things that are important could be known with certainty via logical deduction from self-evident truths. Descartes spoke of "certain laws which God has so established in nature and of certain notions He has impressed within our souls." The mind of man was "tuned" so to speak, to the natural world. The challenge was to discover the laws that govern that world.

Descartes held that the world entailed two basic substances, mind and matter, and like Galileo, he concluded that the natural world was mathematical. His concern was more than trying to provide a framework of thinking for science. He wanted to understand the relationship between

God and His creation—a theological question rather than a scientific question. Did God inhabit the natural world as in the pagan religions? Is God involved "hands-on" in nature where He acts in nature to perform works, which cannot be performed by the laws of nature? Or is God detached from his creation where upon completion of his creative work, He left it to run on its own? These are theological questions but they have important implications for how we should view the natural world and a philosophy of science that "fits" that world.

Accordingly, Descartes' arguments and premises tended to be theological in nature. This perspective was very different from that of Bacon whose focus was on the *mechanics* of the natural world. Descartes' principle argument was the need for a deductive system composed of a consistent set of propositions within which inductive knowledge gained from observations and experiments must fit. Thus certainty of knowledge entailed a blending of these deductive and inductive principles—a reconciliation of Bacon and Descartes so to speak—where that blending was provided through the kind of creative synthesis demonstrated at first by Galileo, who was convinced that the natural world was mathematical, and then by Newton in his development of mathematical physics. This mode of reasoning was formalized by nineteenth century American philosopher Charles S. Peirce[12] as *abduction,* or as it is sometimes referred to as *inference to the best explanation.*

The thinking of Bacon and especially Descartes (and others) led to the *mechanical philosophy,* which viewed the natural world as a machine that operates according to God's natural laws and functions on its own with no need for intervention on the part of God. The God of Descartes was a *watchmaker* God. Ironically, though this view came out of a theological consideration of the relationship between God and His creation, it also provided a rational framework for atheism. Thus, many church fathers, as well as Newton for that matter, were leery of any concept that supported the *mechanical philosophy* because of how easy it was to make the leap from *mechanism* to *atheism*. Nevertheless, in the seventeenth century, Christian natural philosophers were eventually able to reconcile the mechanical philosophy with their understanding of God's rule over nature through natural laws, making the case for a Creator, not atheism. The crucial step in coming to accept this view was the realization that the laws and mechanisms of the mechanical philosophy required a Creator and could not have come into existence through blind chance processes alone. As Bacon noted, take a quick, unexamined look at the world, and you get atheism. Look a little deeper and think a little harder, and you find the Creator. That was true 400 years ago, and it is still true today.

However, today, when it comes to evolution, public education would prefer that students *not* look a little deeper and think a little harder. They might get "confused" by what they learn, and that would really screw up their scores on standardized tests that are based on *not* looking too deep or thinking too hard.

Thus, during this period (sixteenth and seventeenth centuries) and on into the eighteenth and nineteenth centuries there was a general movement in thinking even among Christian philosophers away from *final cause* (an Aristotelian concept) in favor of *mechanism*. The idea was that a God who creates by *mechanism* is greater than a "hands-on" God who has to intervene in His created order to keep everything running smoothly. You might call this the "greater God" hypothesis, which accommodates Darwinism quite nicely and fits the thinking of those who hold to theistic evolution as well as those who hold to atheistic evolution.

Newton, in his thinking (in opposition to Descartes and the vehement objections of Leibniz) supposed that it was necessary for God from time to time to give the planets a little nudge to adjust their orbits to ensure long-term stability of the solar system. While he accepted the mechanical philosophy, he still wanted to believe that God kept his hand in matters just enough for an occasional tune-up. But it was not to be. The great mathematician Laplace, around 100 years later, showed that the gravitational influences of Jupiter and Saturn *ensured* long-term stability of the solar system, giving rise to his famous response to Napoleon's inquiry as to why he had not included any mention of God in his work on celestial mechanics: "I have no need of that hypothesis."

The facts concerning this event are obscure, and it is probably pointless to speculate on whether Laplace was commenting on the existence of God or the non-intervention of God in his creation. There is ample room to spin this story any way you want. What is perhaps more significant than Laplace's opinion about God's role in nature (whatever it might have been) was his demonstration mathematically of how the "just-right" orbital parameters of Jupiter and Saturn provided long-term stability to the solar system, one of the many "just-right" anthropocentric conditions necessary for a suitable habitat for mankind.

❧ ❧ ❧

A general philosophy of nature gradually took shape out of the debates among natural philosophers during the seventeenth and eighteenth centuries. According to this view, God, in creating the natural world, put in place natural laws and principles and ordered the world in all of the intricate details necessary for a universe, which provides not only a habitat for living creatures but the "stuff" that makes up the living creatures themselves. In *this* universe, which operated fully within the constraints of the mechanical philosophy, God used natural laws and processes to create life and mankind, thus establishing the mindset in the nineteenth century for the search for the origins of life and mankind through *natural* mechanisms. While this search for an understanding of the origins of man may have started out within the Christian view of the world, it would wind up in the naturalistic view of the world.

This understanding of nature is described by Cornelius Hunter in his book *Science's Blind Spot*[13] as *theological naturalism* because it was formulated from *theological* considerations. This view entailed the "greater God" hypothesis, which is no different mechanically than the "no God" hypothesis. Regardless of your preferred metaphysical view, "greater God" or "no God," with respect to its mechanics and day-to-day operations, you looked at the natural world in the same way. As for the origins of the universe, of life, and of mankind, this was another matter. While the principles of methodological naturalism were conceived, at least in principle, within a Christian view of the world over 300 years ago, they applied only to the mechanical aspects of the natural world, not its origins. As to whether or not life could be explained through natural mechanism, that remained to be seen. Consequently, the legitimate domain of methodological naturalism reaches *only* to the mechanics and operations of the natural world as described within the *mechanical philosophy*.

Another concept that grew out of this period was that of *reductionism*, where it was held by naturalists that if you could reduce everything in nature to its fundamental parts and then examine the interactions between those parts, you would find that there was no place or requirement for God and that all natural phenomena could be explained without appeal to the supernatural. Given that naturalists also viewed life as a strictly natural phenomenon, it seemed appropriate that reductionism should be applied to life as well and with the same expectation concerning God.

Reductionism seemed to have some merit within mechanics, but as discovered in the twentieth century, the reductionist approach to living creatures was a different matter. Eighteenth- and nineteenth-century naturalists had no idea as to the staggering complexity of the biology cell and the processes that took place there.

≪ ≪ ≪

The concepts of law and design in nature were presupposed in the doctrine of creation, which viewed God as Lawgiver and Designer. Consequently, it was natural that both concepts would be incorporated in developing a philosophy of science. Further, there was complete harmony with what was observed in terms of the ontological properties of the natural world (a beginning, order, design, purpose, law, and principle) and the doctrine of creation as understood within the Christian view of the world. At this point in the history of science, the question regarding the origins of life was an open matter, at least with respect to the details. Then, as well as now, the means by which life came into existence was, and is, unknown scientifically. While the "greater God" hypothesis and the ideas of "creation by mechanism" left the door open to some mechanical or naturalistic means for the creation of life (such as chemical evolution), such thinking was based on a rudimentary and vastly incomplete understanding of biology. It is apparent in light of present knowledge in the biological sciences that when it comes to giving an account of how living creatures came into existence, creation by mechanism no longer appears to be a reasonable option.

◈ ◈ ◈

The great *metaphysical* questions that underlie science are: What were the causes of the universe, its order, its governing laws and principles, and of life? It is unavoidable that the natural sciences will touch on such questions. If the material world was created by a transcendent intelligence, was designed for a purpose, and has some ultimate meaning, then as rational creatures, we might expect to see positive evidence of such in the material world. If the material world was not created and is the result of unguided material causes alone and we are simply accidents of nature, then since such causes have no innate capacity for producing a world with meaning and purpose, we might expect to find no evidence for such.

But we have reasoned ourselves into a profound dilemma here. If naturalism is true and the natural world has no meaning, we could never discover that it has no meaning. As C. S. Lewis noted in *Mere Christianity*,[14] "If the whole universe has no meaning, we should never have found out that it has no meaning; just as, if there were no light in the universe and therefore no creatures with eyes, we should never know that it was dark. Dark would be a word without meaning."

If naturalism is true, how could we know it is true? Naturalism would be a word without meaning, yet, here we are talking about it as if it has meaning. Lewis' reasoning compels us to conclude that naturalism is a *meaningful* concept and therefore cannot be true.

− − −

A major objective here was to introduce some of the conceptual challenges early natural philosophers faced in developing what was both a practical *methodology* for scientific investigation, a *way of thinking* about the natural world and an understanding of the place of God in the natural sciences. We have also shown that methodological naturalism actually had its roots in Christian theological considerations concerning the nature of the created order and God's role in the day-to-day workings of the natural order.

What has been addressed here hardly tells the whole story. In fact, more of that story will be addressed in chapter 4 where we reveal the true historical relationship between science and Christianity, in chapter 6 where we examine what is meant by "natural causes," and in chapter 9 where we address the works of three eighteenth-century philosophers, David Hume, Immanuel Kant and William Paley, all of whose works strongly influenced Darwin's thinking as he developed his theory of biological evolution. With the exception of Paley, their works have also been used, albeit through misrepresentation, to construct arguments against the teleological view of nature.

CHAPTER 4

THE "WAR" BETWEEN SCIENCE AND RELIGION

I n this chapter, we dispose of a commonly accepted myth about the relationship between science and religion that is used to protect the exclusive status of Darwinism in public education. This myth allows Darwinists to accuse anyone who might criticize Darwinism (regardless of the legitimacy of the critique) of being religiously motivated. There is no point in taking such criticism seriously since everyone "knows" (according to the myth) that science and religion are two fundamentally incompatible ways of looking at the world, and it would be wrong to try to relate scientific thinking in any way to religion. Everyone, including the courts, seems to buy this. An honest look at the history of science, however, shows that this notion is inherently bogus. Professor of the history of science and technology at Johns Hopkins University, Lawrence M. Principe, provides an excellent overview of this mythology in his audio CD produced by the Teaching Company,[1] adamantly making the point that no historian of science today accepts the warfare thesis.

The modern popular view about the relationship between science and religion is that these two great areas of human experience are fundamentally at odds with each other. They are in conflict, they have always been in conflict, and they always will be in conflict. It is inherent to their natures. To fully understand this relationship, however, we must consider the full framework of relationships between science, naturalism, and Christianity as accepted today by the academic world. That framework is shown in Figure 4.1. Here, we refer specifically to "Christianity" rather than "religion" because of the unique role Christianity played in the historical and philosophical development of science, a role not shared with any other religion. Further, we introduce the term *conflict* along with *war* to more properly characterize the nature of these relationships. The "war" between Christianity and naturalism is philosophical in nature, having to do with the nature of reality. It is unavoidable that science will be in conflict with one and in accord with the other. Which has a legitimate historical and rational claim on science?

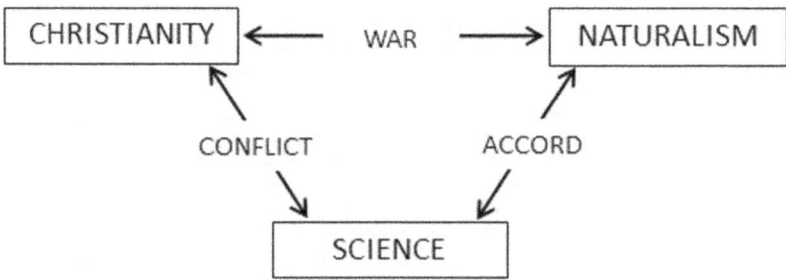

Figure 4.1   The relationships between science, naturalism, and Christianity

as accepted today in American culture and institutions

(but not by historians of science).

In Figure 4.1 we see *naturalism* and *science* in one accord in their alignment against *Christianity*. While this is the view that is commonly held today in popular culture, it has nothing to do with the actual history of these relationships. In reality, it was the Christian doctrine of a created order that provided the presuppositions necessary for science to develop. The warfare myth (or "conflict" myth, as used in Figure 4.1) as in the case of most mythologies was established through the process of revising history and then disbursing that revised version of history throughout the culture. How? By introducing it in professional writings and then the curriculum of our educational systems—first in the universities and then in public schools. While it took several generations to fully establish the warfare myth, it was in effect a bloodless revolution from the inside out, and the best part (at least from the point of view of the naturalist) was that it was funded by the parents of those children targeted for indoctrination. For anyone interested in Marxist-style cultural revolutions, one could not imagine a sweeter deal. Gain control of public education, and you gain control of the minds of future generations. You own the future without firing a shot!

We address the Galileo affair, the writings of two late nineteenth century authors named Draper and White, and the Scopes trial to provide historical context for today's warfare debate. We take a brief look at actual historical scholarship, which exposes the warfare myth, consider the views of some of the great physicists of early twentieth century concerning the relationship between science and God, and then make some final observations in which we paint a more complete picture of the relationships between science, Christianity and naturalism.

The Galileo Affair

The "Galileo affair"[2] in the early seventeenth century is often portrayed as the event that started the "war" between science and religion. In the popular account, Galileo, the humble truth-seeking scientist, is brought before the Inquisition, tortured, and imprisoned by a dogmatic power-hungry Catholic church because his heretical Copernican (heliocentric) views challenged the geocentric dogma of the church.

This makes a great story, and it is really effective in stirring up antagonism toward that mean old authoritarian Catholic church and silencing Bible-thumpers, but it is largely a fabrication. The fact is that it was the philosophers and the professors (the scholastics) in the universities who were the defenders of Aristotle's geocentric views. Consequently, they, along with certain Dominicans within the church, who were more concerned with internal church politics than science, were the ones who condemned Galileo for his adventures into Copernicanism.

There were in fact *two* conflicts, and Galileo was at the center of both. The first was with the scholastics in the universities and was scientific in nature, having to do with the structure and dynamics of the universe— at least as they understood it at that time. The second had to do with Augustinian teachings about the rules for scriptural interpretation. There was also a personal twist to this second conflict because it would eventually involve Galileo's old friend Cardinal Maffeo Barberini.

And then there was the "Galileo factor." Galileo had issues. He was brilliant and admired, but he was also arrogant and had poor social skills, especially regarding those who criticized his views. Consequently, he effortlessly made enemies and often for no good reason. It is almost certain that if Galileo had exercised a reasonable amount of humility and diplomacy and a lot less arrogance, events would have played out to a conclusion far more to his liking.

The church was largely ambivalent about the Copernican theory itself, where it was regarded as merely a "mathematical device" with nothing to say about the nature of reality. It was instrumental or useful, not descriptive or doctrinaire. That was true in the days of Copernicus, and it was also true in the days of Galileo. With respect to such scientific matters, the principle concern of the church was scriptural interpretation, not science itself. The church had little concern about mathematical devices so long as they were *only* mathematical devices. When Copernicus' book was published in 1542, it was greeted with little controversy. Nevertheless, the fires of controversy were smoldering, and eventually in 1616, his book was placed on the index of forbidden books, probably because of the increased attention stirred up by Galileo's interest in this subject, which in turn, caught the attention of the scholastics and the church.

Another factor that is almost always lost in any discussion of the Galileo affair is the extent of authority held by the Catholic church in the sixteenth and seventeenth centuries, where there was a consolidation of the secular and the sacred under church authority. Whenever there was a secular issue—whether it had to do with civil government or engaging in war or scientific discovery—if it touched on church authority, then it was a church matter, and such was the case when Galileo proposed to write a book on the Copernican system. Thus it was this overarching authority of the church, concerning civil affairs, which set the stage for the Galilean drama.

The principles of scriptural interpretation concerning scientific find-ings that appeared to conflict with accepted scriptural interpretation were first articulated by St. Augustine of Hippo (AD 354–430). Augustine said that if an occasion should arise where scripture and conclusive scientific findings were clearly in disagreement, then a formal process should be instigated to reconcile scriptural interpretation to plain facts derived from scientific findings. Augustine argued that all truth is God's truth, and truth cannot contradict truth. It is error in human understanding that creates such apparent contradictions. In some cases reconciliation might be achieved by simply recognizing that a particular scripture should be taken *metaphorically* rather than *literally,* or perhaps a true understanding of a certain scripture requires a more sophisticated contextual analysis of relevant historical or cultural factors. Galileo understood these principles as well as any of the leaders within the Catholic church.

If such an occasion arose, the first step would be to subject the relevant scientific findings to meticulous scientific scrutiny to firmly establish their credibility. Once that credibility was authenticated, the church would take action to reconcile scriptural interpretation to scientific fact. To ignore these principles was to invite heresy, and this is where Galileo ran afoul of the Catholic church.

We must also remember the state of scientific knowledge at that time; things that we accept today as rudimentary knowledge were, in Galileo's time, great mysteries that had not yet been resolved. Likewise, the church was being challenged, not just with new ideas about the structure of the universe, but with how to relate to knowledge about God's creation obtained through secular means. Historically the church was the defender and protector of revealed truth, but who gets to decide what is or is not true regarding scientific knowledge, especially when that knowledge has something to say about deeply religious matters, such as the origins of the universe and the nature of reality?

Then there was the problem of Aristotle. Thomas Aquinas had achieved a general reconciliation between Catholic doctrines and Aristotle several hundred years before, thus further embedding Aristotelian philosophy in church thinking. Now, the foundations of Aristotelian cosmology were being shaken by the natural philosophers of the sixteenth and seventeenth centuries, and if Aristotle's foundations were shaking, then you can be sure that there was some shaking going on within the church as well.

In 1623 Cardinal Maffeo Barberini was elected Pope, taking the name Urban VIII. A year later, Galileo visited his old friend in Rome where he was warmly received. During this visit, Urban encouraged and blessed Galileo's plans to publish a book entitled *Dialogue Concerning the Two Principal Systems of the World*—those *systems* being geocentrism and heliocentrism. Urban gave his approval for publication with the understanding that because of the implications of the heliocentric view for scriptural interpretation and lack of conclusive scientific evidence in support of that view (and he was right about that), Galileo should write *hypothetically*—a perfectly reasonable request. Galileo had this same discussion with Cardinal Roberto Bellarmino fourteen years before and agreed in writing as a matter of official record that he would treat the Copernican theory hypothetically.

Having agreed to the terms for publication, Urban assured Galileo that as long as he was pope, he would have his protection from the powerful university professors as well as critics within the Catholic church, which was never lacking in internal drama. Galileo returned to Florence, finished his book, and then published it. Then all hell broke loose.

Two things got Galileo in trouble. First, he did not write *hypothetically.* Instead he defended the Copernican theory as fact when there simply was no legitimate scientific basis for such a defense. The second problem was Galileo's astonishing personal insult of the pope in the very book the pope himself had so graciously encouraged Galileo to write. Galileo argued that the tides were caused by the earth's motion around the sun—like water sloshing in a bucket—and that this *proved* the heliocentric view. Urban suggested that Galileo include a statement in his book to the effect that God, through his unlimited power, could use other means to account for the motion of the tides. Galileo agreed that he would include such a statement in his book and he did.

Galileo wrote his book as a dialogue between three characters. The hero of the dialogue was Salviati who made Galileo's arguments for him in support of the Copernican system. The impartial mediator of the dialogue was Sagredo, and it was his job to ask the questions and guide the discussion. The third party in the dialogue was a buffoon sort of character named Simplicio who defended Aristotle's geocentricism with absurd arguments. So, which of these characters will Galileo have to represent the pope—Salviati, Sagredo, or Simplicio?

On the last page of his book, Galileo put the words of Urban in the mouth of—not Sagredo and not Salviati—but Simplicio. Having spoken Urban's words, Simplicio (the pope) was then mercilessly mocked by Salviati (Galileo) for saying such an absurd thing. None of this escaped Urban. He was outraged and had Galileo brought before the Inquisition.

Had Galileo put the words of Urban in the mouth of Salviati—a most reasonable thing to do where it would have been treated with the respect it deserved—then the fact that Galileo failed to write hypothetically would likely have been handled by the Inquisitors (assuming it was even bought before the Inquisition) at the level of a preliminary investigation with perhaps some stern language about Galileo getting carried away with his argument for Copernicanism, followed by a slap on the wrist and then be done with it. But this was a very bad year for the pope in that he had to deal with troublesome things such as the Thirty-Year War in Germany that was going poorly, the plague wreaking havoc in Italy, and then the movement in Spain to have him impeached because, among other things, he was "soft" on treatment of heresies—which was precisely the issue that Galileo dropped into Urban's punchbowl.

Galileo spent the remaining years of his life under house arrest in his own villa near Florence where he lived comfortably and received many friends, without imprisonment, torture, or dungeons. It was during this time in his life that he published his seminal work on *Mechanics* and *Strength of Materials* that did so much in preparing the way for Newton. Scientifically, this work was to be Galileo's great legacy—not heliocentrism. He also played a pivotal role in making the final transition in the philosophy of science from the search for Aristotle's obscure *causes* to the search for *mechanism and law.*

Later, Newton showed that it was the gravitational influence of the sun and the moon that accounted for the tides. As it turned out, on this point, the pope (Simplicio) was right, and Galileo (Salviati) was wrong.

Galileo certainly did not deserve the punishment he received, but he could have spared himself a lot of trouble with a little humility and common sense. Keeping his word and not embarrassing the pope would have radically changed the outcome, but then he may never have done his seminal work on mechanics. It's interesting the way things turned out: a happy ending.

Draper and White

The origin of the modern mythology of *the war between science and religion* can be traced back to the work of two authors who wrote during the later decades of the nineteenth century, John William Draper and Andrew Dixon White.

John William Draper was the first president of the American Chemical Society, and in 1874, wrote a book entitled *A History of the Conflict between Religion and Science*.[3] His sentiments are summarized in the first few paragraphs of his book. "The history of science is a narrative of the conflict of two contending powers; the expansive force of human intellect [science] and the compression arising from traditional faith [religion]." In short the message of history, according to Draper, is that science is good for mankind, and religion is bad for mankind. His book was essentially one long diatribe against the Catholic church, reflecting his own personal anti-Catholic views.

Andrew Dixon White was first president of Cornell University, the first secular university founded in the United States, and in 1896, he wrote *A History of the Warfare of Science with Theology in Christendom*.[4] Unlike Draper, White was actually a historian, and in his writings, he was somewhat more scholarly and considerably less rabid than Draper. It appears that the motivation for his book, at least in part, was retaliation against

Christian academics for the criticism leveled at him for establishing a purely secular university.

The underlying premise of both books is that science and religion have always been in conflict, where throughout history, religion has had an oppressive influence on scientific development and the expansion of human knowledge. Both books had deep underlying political and anti-religious motivations. They were filled with distortions of facts, quotes out of context, and outlandish generalizations. Both books blatantly misrepresented the true history of that which they claimed to be giving an account.

These books appeared at a time in American history when the universities were undergoing radical changes from primarily a Christian orientation to the secular German progressive model of education. The writings of Draper and White also fit nicely into the revolutionary views that were being implemented in America's universities during this period, and both faculty and students dutifully drank the Draper/White Kool-Aid. By mid-twentieth century, the mythology was solidly entrenched in America's academic and cultural worlds.

These were terrible books that effectively established the mythologies that inform opinions today across our culture. Even though they were based on appalling scholarship, their message was nevertheless quickly embraced by students and faculty alike because the minds of the students and faculty had been preconditioned in part by another mythology manufactured by extreme Enlightenment philosophers in what is called the *Dark Ages*.

During the last half of the twentieth century, scholars began revisiting the history of the West and discovered that the *Dark Ages* was a fabrication. Rodney Stark, now at Baylor University, is one such scholar, and some of

57

his comments regarding the Dark Ages give critical perspective to Draper and White. In his book *The Victory of Reason; How Christianity Led to Freedom, Capitalism, and Western Success*,[5] Stark says:

> For the past two or three centuries, every educated person has known that from the fall of Rome until about the fifteenth century Europe was submerged in the "Dark Ages"–centuries of ignorance, superstition, and misery–from which it was suddenly, almost miraculously rescued, first by the Renaissance and then by the Enlightenment. But it didn't happen that way. Instead, during the so-called Dark Ages, European technology and science overtook and surpassed the rest of the world.

> The idea that Europe fell into the Dark Ages is a hoax originated by antireligious, and bitterly anti-Catholic, eighteenth-century intellectuals who were determined to assert the cultural superiority of their own time and who boosted their claim by denigrating previous centuries as—in the words of Voltaire—"a time when barbarism, superstition, [and] ignorance covered the face of the world."

Elsewhere Stark notes that the *Renaissance*, another creation of the Enlightenment, actually started, not with the re-introduction of classical Greek learning into the West in the fourteenth century as traditionally held, but when Rome fell, and the single biggest factor that sparked that renaissance was that when Rome fell, its brutal system of taxation fell with it. When the suffocating burden of Roman taxes was lifted, hope

58

and optimism returned to the common folk throughout Europe. A new spirit swept across Europe, which rapidly developed a technological lead over the rest of the world. This technological advantage would come in handy several hundred years later when southern European armies used advanced military technology to decisively defeat the invading Muslim armies, literally saving Western civilization!

The Scopes Trial

In late nineteenth century in America, evolution faced legitimate scientific challenges. There were doubts among scientists regarding the creative abilities of natural selection and evolution lacked a theory of inheritance to preserve favorable variations. Many entertained the concept of evolution and the idea that it might be guided, but not by natural selection. Princeton theologian Charles Hodge, in his influential book (1874) *What is Darwinism?* said, "It is atheism . . . denial of design in nature is virtually the denial of God." Many agreed and still do.

Harvard botanist Asa Gray, however, a devout Protestant, viewed evolution as a "higher . . . worthier . . . more consistent . . . view of design in nature than heretofore." He clearly supported the "greater God" hypothesis and the idea of creation by law.

Evolution did not become a fighting matter for many ordinary folks until it began to influence their children's education. In the 1920s, the majority of southern legislatures *rejected*—I repeat—*rejected* antievolution laws. Almost all biology teachers in high schools and colleges believed in some form of biological development. The Butler Act passed in January 1925 by the Tennessee House of Representatives was more the exception (along with Oklahoma and Florida) than the rule. The Butler Act held that it was unlawful for state-supported schools *"to teach any theory that denies the story of the Divine Creation of man as taught in the Bible, and to teach instead that man has descended from a lower order of animals."* Six weeks later the Tennessee Senate passed the bill that was then signed into law by Governor Austin Peay.

The newly formed ACLU in New York offered to defend any Tennessee teacher who was accused of teaching evolution in defiance of the Butler Act as a test of its constitutionality. A local Dayton businessman, the county school superintendent, and a local lawyer agreed that the controversy of such a trial would attract a lot of attention to Dayton. They approached John Scopes (who wasn't even a science teacher), and a deal was made.

William Jennings Bryan, three-time Democratic presidential candidate and leader of the fundamentalist's efforts to outlaw the teaching of *human* evolution joined the prosecution. Clarence Darrow, famous Chicago lawyer and agnostic (and all-around despicable human being), joined the defense. Among great hype and fanfare, the "trial of the century" was on.

Bryan's mission in life was defending Christian faith against evolution. Darrow's mission in life was to rid the world of Christianity. In fact, he cared far more about condemning Christianity than defending evolution. Bryan argued, "The real issue is not what can be taught in public schools but who shall control the educational system? If the people are not to control the schools, who shall control them?" Scientists? Teachers? Politicians? Bryan was a champion of the common people. Darrow's strategy: Put Bryan on trial, not Scopes, and argue for academic freedom in public education.

Dayton civic leaders wanted a public relations event. They got it. Bryan wanted to defend the faith and the interest of the people. He did. Darrow wanted to flail against Bryan and Christianity. He did. The ACLU wanted a shot at showing that the Butler Act was unconstitutional. Their interest was washed away in the melee.

Bryan won. Scopes was convicted and fined $100 (a good chunk of money in those days), which Bryan agreed to pay. After a few days of trial, even the defense began to tire of Darrow's endless rants about Christianity and his attacks on Bryan. By the end of the trial, everyone had had enough of Darrow. Bryan walked away a hero to the common folk. Sadly, a few days later, he died in his sleep during an afternoon nap. No doubt the heat and the stress of the Dayton event had taken its toll.

Unfortunately, the account that persists in modern folklore has little to do with the actual history of the Scopes Trial where that history was retold in a book, a play, and a movie.

In his book *Only Yesterday* (1931), Frederick Allen reduces Christian fundamentalism to antievolutionism and antievolutionism to Bryan. In Allen's account, Bryan is completely humiliated, crushed. It is a decisive defeat of "old-time religion," a triumph of reason over revelation, and the triumph of science over religion. Allen's book was actually used in history classes in universities as if it had historical merit.

This scenario was repeated in 1960 in the movie *Inherit the Wind* where Bryan is transformed into a mindless hypocrite who attacks all science as godless. The town of Dayton takes on a sinister mood when, in fact, the people of Dayton were well-behaved, enjoyed the spectacle, and in typical Southern tradition were quite hospitable to the many visitors. Darrow is shown to be tolerant of Bryan's beliefs—foolish as they were.

Inherit the Wind was severely criticized in published reviews for its terrible distortions of the facts, but it was the myth that lived on in American culture—and still does.

Edward J. Larson tells the whole fascinating story, from Dayton to Hollywood, in his wonderful Pulitzer Prize–winning book, *Summer for the Gods*.[6] If you want to understand the truth about what happened at Dayton that summer in 1925, read this book. It's a great read! Larson knows how to tell a story.

God and Nature

The warfare thesis has been completely discredited by modern scholarship where no serious historian of science supports that thesis—*not one!* Of particular note is the publication in 1986 of a collection of eighteen scholarly essays written by an international group of distinguished historians entitled *GOD AND NATURE; Historical Essays on the Encounter between Christianity and Science*.[7] This landmark work by historians David C. Lindberg and Ronald L. Numbers "promises not only to silence the persistent rumors of war between Christianity and science, but also to serve as the point of departure for new explorations of their relationship."

It has been thirty years since these essays were published. More recent scholarly works have appeared, such as "Science and Religion: A Historical Introduction"[8] edited by Gary B. Ferngren and "When Science & Christianity Meet"[9], another collection of essays edited by Lindberg and Numbers. Little has changed. The warfare thesis is still widely accepted as historical truth.

The Physicists

It is not widely known today, but many (certainly not all, Niels Bohr being a notable exception) of the physicists of early twentieth century who laid the foundations of our modern understanding of relativity and quantum mechanics, openly spoke of a deep conviction as to the reality of God and their belief in an eternal sovereign Creator and Ruler over the universe who was due the adoration and worship of man. We look at the testimonies of Max Planck, Albert Einstein, and Werner Heisenberg to get a sense of the depth and sincerity of their beliefs.[10]

In his famous lecture "Religion and Science" (May 1937), Max Planck (Nobel Prize in Physics, 1918) said:

> Both religion and science need for their activities the belief in God ... It is the steady, ongoing, never-slackening fight against skepticism and dogmatism, against unbelief and super-stition, which religion and science wage together. The directing watchword in this struggle runs from the remotest past to the distant future: "On to God!"

While Albert Einstein (Nobel Prize in Physics, 1921) certainly did not hold to traditional ideas about a personal God, he had much to say about faith and science that reflects a deep devotion and reverence for "his" God. Here is a popular quote, just one of many, reflecting his thoughts about physics and God:

63

I want to know how God created this world. I am not interested in this or that phenomenon, in the spectrum of this or that element. I want to know His thoughts, the rest are details.

Werner Heisenberg (Nobel Prize in Physics, 1932) said:

In the history of science, ever since the famous trial of Galileo, it has repeatedly been claimed that scientific truth cannot be reconciled with the religious interpretation of the world. Although I am now convinced that scientific truth is unassailable in its own field, I have never found it possible to dismiss the content of religious thinking as simply part of an outmoded phase in the consciousness of mankind, a part we shall have to give up from now on. Thus in the course of my life I have repeatedly been compelled to ponder on the relationship of these two regions of thought, for I have never been able to doubt the reality of that to which they point.

Beyond Myth

Draper and White created the twentieth-century myth of the war between science and religion. What actual history teaches us is that science and Christianity are in deep accord and partners in their quest for knowledge of the truth and an understanding of the natural world. Four hundred years ago, science and Christianity opened up this chapter in the affairs of mankind, and there is no reason why we should not expect that together they will write more chapters in the future.

Those who read Draper and White failed to understand these things because their minds were predisposed (*poisoned* might be a better word) toward the new European secular and progressive ideals that were in the air at the turn of the twentieth century, which, inspired by the *mechanical philosophy,* saw less and less of a role for God in the affairs of men. Once the warfare mythology was set loose on the academic world, it was quite natural to incorporate into that mythology distorted versions of other historical events such as the Galileo affair and the Scopes trial, thus adding a look of authenticity to pure fabrication. To further exasperate the situation, this mythological war ultimately led to a "war in fact" where many in the Christian community developed a sense of distrust of the scientific community—and rightfully so—because naturalistic scientists were using it to challenge the fundamental premise of theistic faith: the creation by God of the universe and of mankind. The scientific community responded with contempt, and it wasn't long until mythology had become "reality."

⋆ ⋆ ⋆

We have clearly shown here that science and Christianity are not in conflict either historically or with respect to fundamental presuppositions. In fact, they are in one accord. The truth is that in this conflict, Christianity and science are aligned in one accord against naturalism.

Philosophers of Science Michael Ruse and Alvin Plantinga, to their immense credit, make this point. Ruse in his book *The Creation Struggle,*[11] defends naturalistic science but is honest enough to admit that "we have no simple clash between science and religion but rather between two religions." Plantinga, in his book *Where the Conflict Really Lies,*[12] concludes that "there is indeed a science/religion conflict, all right, but it is not between science and theistic religion, it is between science and naturalism. That's where the conflict really lies." Ruse concludes that the conflict is between naturalism and Christianity. Platinga concludes that the conflict is between science and naturalism. Both are right. Put them together and you get the full story. This correct arrangement of relationships is presented in Figure 4.2 and shows that naturalism is the enemy of both Christianity and science. This, incidentally, is precisely the point made by both Max Planck and Werner Heisenberg.

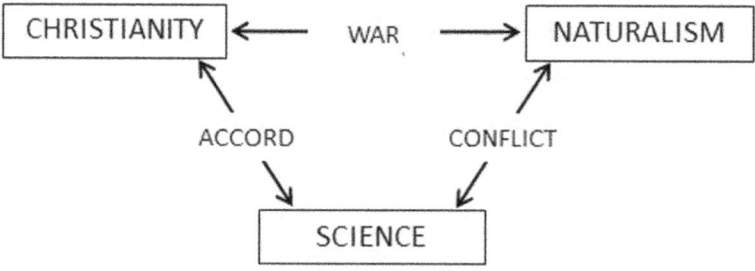

Figure 4.2   The true relationships between science, Christianity, and

naturalism as reflected in history and in principle.

The great power of myth is that it is founded, disseminated, and preserved where ignorance and prejudice prevail within the culture and is almost always posed within the context of a simplistic "good guy–bad guy" scenario. In the expression of this mythology in the current debate between Darwin and design, Darwin (the good guy) is cast as *science,* design (the bad guy) is cast as *religion*, and science and religion are cast as relentless enemies in the battle between truth and superstition. As long as this mythology finds general acceptance throughout the principle institutions of our culture (especially in public education and in the judiciary), scientific evidence, historical fact, and rational arguments will be irrelevant. Only truth can free us from the tyranny of mythology. The process of such liberation is particularly difficult when it concerns institutions like public education, the courts, and the media where myths, like Bruce Willis, "die hard."

> *No modern historian of science supports the warfare thesis.*
> *Historically, science is recognized more as a legacy of*
> *Christianity than an enemy of religion.*

CHAPTER 5

SCIENCE AND NATURALISM

P rior to the twentieth century, the great majority of natural philoso-
phers/scientists were Christian. That is no longer the case, at least
for those who run the scientific establishment. The majority of members
of the National Academy of Sciences, university science department
heads, journal editors, presidents of professional scientific associations,
and senior scientists at the National Institute of Health and the National
Science Foundation, collectively (and unofficially) referred to as the "scientific priesthood," are generally agnostic or atheistic in their beliefs. The
presence of a committed evangelical Christian within this "priesthood" is
generally regarded as an anomaly and in some cases, an embarrassment.
Larson and Witham[1] showed in a survey conducted in 1997 that 93 percent
of the members of the National Academy of Sciences were atheists or
agnostics, and there is no compelling reason to believe that the numbers
would be substantially different for the upper levels of the scientific
establishment in general. Historical trends would suggest that this number
might be even higher today, twenty years later.

Thus, there exists a strong predisposition within today's scientific community, especially in biology, toward naturalism. What happened to cause this dramatic shift within science from a predominately Christian view of the world to a predominately naturalistic view of the world and in only 100 years? While there were a number of cultural factors that contributed to this revolution (an important one being the liberalization of Christianity in Europe in the nineteenth century), it was Darwinism. By mid-twentieth century, Europe had essentially fallen to Darwinism. America, however, has proven to be a real challenge where overcoming the influence of conservative Christianity as a dominant cultural and intellectual force is proving to be a difficult, if not hopeless, task. Major battlegrounds in this struggle are public education and the courts.

Consider some statements made by leading naturalistic scientists that reflect both their naturalistic vision and the depth of their naturalistic commitment.

Douglas Futuyma[2] (Stoney Brook University):

> By coupling undirected, purposeless variation to the blind, uncaring process of natural selection, Darwin made theological or spiritual explanations of the life processes superfluous. Together with Marx's materialistic theory of history and society and Freud's attribution of human behavior to influences over which we have little control, Darwin's theory of evolution was a crucial plank in the platform of mechanism and material-ism—-of much of science, in short—-that has since been the stage of most Western thought.

John Searle[3] (University of California Berkeley):

Acceptance of the current views is motivated not so much by the independent conviction of their truth as by the terror of what are apparently the only alternatives. That is, the choice we are tacitly presented with is between a "scientific" approach, as represented by one or another of the current versions of materialism, and an "unscientific" approach, as represented by Cartesianism [mind-body dualism] or some other traditional religious conception of the mind.

Franklin Harold[4] (Colorado State University, University of Washington):

We should reject, as a matter of principle, the substitution of intelligent design for the dialogue of chance and necessity ... but we must concede that there are presently no detailed Darwinian accounts of the evolution of any biochemical system, only a variety of wishful speculations.

Will Provine[5] (Cornell University, deceased 2015):

Evolution is the greatest engine of atheism ever invented.

Daniel Dennett[6] (Tufts University):

Darwinism is . . . a universal acid: it eats through just about every traditional concept and leaves in its wake a revolutionized world-view.

Jacques Monod[7] (Institut Paris, deceased 1976):

The ancient covenant is in pieces. Man at last knows that he is alone in the unfeeling immensity of the universe, out of which he emerged only by chance.

Publications (available online) by the National Academy Press, Washington, DC, further reflect the naturalism that is so deeply entrenched in the scientific establishment and concealed in science education in public schools. Such publications include: (1) a booklet defending evolution against creationism: *Science, Evolution and Creationism* (2008); (2) the *National Science Education Standards* (1996) developed as a model for science education for all states; (3) more recently the *Next Generation Science Standards* (2013) which as of November 2017 had been adopted by nineteen states; and (4) a guideline for teaching evolution in public schools *Teaching about Evolution and the Nature of Science* (1998), which is little more than an apologetic for naturalism.

ৰ্৯ ৰ্৯ ৰ্৯

Benjamin Wiker in his book *Moral Darwinism* traces the roots of naturalism in Western civilization back to the Greek philosopher Epicurus (341–270 BC). What follows here is basically a brief (very brief) collection of excerpts from Wiker that summarize the impact of Epicurus on the history of the West. Here is Wiker[8]:

Epicurus sought a practical guide to happiness. True pleasure rather than absolute truth was his objective. Experience rather than principle was his guide. He warned against the excesses of Hedonism (physical pleasure is the highest good) and advises us to be content with the simple things. Some pleasures were higher than others—such as friendship which was valued above all the rest.

Epicurus' intentions went far beyond pleasure as an end in itself. He sought something deeper, a system of thought, an approach to life, a worldview that not only exalts pleasure but also provides peace of mind and serenity. What good is pleasure if you are always troubled? This state of serenity, Epicurus concluded, can only be achieved by finding a remedy for the unsettling effects of a troubled conscience in this life and fear of judgment in the next. What was needed was a philosophical system that had no place for either moral law or divine judgment.

Epicurus dispenses with the discomfort of guilt and fear of judgment by embracing a view of nature that has no place for God. With no place for God, neither can there be a place for divine morality or judgment.

How then does one pursue such a philosophy? How does one live and think if he is to enjoy the fruits of Epicureanism? Epicurus points the way. It is through "right thinking" and "habits of mind" which view nature as a closed system functioning independent of the influence of God. Further, he says, the study of the natural sciences aided by the light of "right thinking" will reveal that the natural world is in fact self-existing with no God, no moral law and no judgment. Through this means one can become a fulfilled Epicurean.

One could hardly fashion a better summary of the view of modern naturalistic scientists. In fact as we shall see in chapter 12, Epicurus may have been the first *positivist*.

Under the influence of the Roman poet-philosopher Lucretius (99–55 BC), epicurean materialism was well established in the Greco-Roman world by the time of the birth of Christ. Its principle competitor was stoicism which embraced the teleological view of the natural order, acknowledged a universal moral law (also called natural law) given by God and held virtue, not pleasure, as the highest good. It is interesting that while the Hebrews received the Mosaic law as revealed by God, the stoics received natural law as revealed by nature. Both in their own way prepared the way for Christianity, one through special revelation and the other through nature.

It was these two philosophical systems, Epicureanism and stoicism, that the apostle Paul encountered at Athens on Mars Hill as recorded in the seventeenth chapter of Acts. The Mars Hill event was indeed a historic moment.

୶ ୶ ୶

As evidenced by the writings of the Church Fathers during the first four hundred years after Christ, epicureanism did not embrace a "live and let live" philosophy (neither do its heirs embrace such a philosophy today) in that Christianity and epicureanism were natural and bitter enemies. By the time Rome fell during the fifth century AD, epicureanism had succumbed to the power of the Gospel of Christ and was effectively dead as a cultural and spiritual force. It looked like it was all over for Epicurus.

For 1000 years, Epicurus "slept." But suddenly, through the rediscovery of his writings by European scholastics during the Renaissance, he was awakened to a new world. The world had dramatically changed, and those changes spelled new hope for Epicurus. The *mechanistic* view of the cosmos, which drew its reasoning from Newton's laws of mechanics, breathed new life into Epicurus' materialistic philosophy. Naturalism as a worldview was reenergized and with the introduction of Darwin's theory of evolution in the last half of the nineteenth century, epicureanism was ready to resume battle with its ancient foe. Its mission: the utter destruction of Christianity, root and branch. Epicurus may have lost Round One, but now it was time to engage in Round Two. Fight on!

Evidence of this epicurean mission is reflected broadly in atheistic publications—sometimes subtly and sometimes not so subtly as, for example, in his article "The Meaning of Evolution" in the *American Atheist*,[9] Richard Bozarth says:

> Atheism is science's natural ally. Atheism is the philosophy, both moral and ethical, most perfectly suited for a

74

scientific civilization. If we work for the American Atheists today, Atheism will be ready to fill the void of Christianity's demise when science and evolution triumph. Without a doubt humans and civilization are in sore need of the intellectual cleanness and mental health of atheism.

Christianity has fought, still fights, and will fight science to the desperate end over evolution, because evolution destroys utterly and finally the very reason Jesus' earthly life was supposedly made necessary. Destroy Adam and Eve and the original sin, and in the rubble you will find the sorry remains of the son of god. Take away the meaning of his death. If Jesus was not the redeemer who died for our sins, and this is what evolution means, then Christianity is nothing!

Another kindly thought from John J. Dunphy in his article "A Religion for a New Age," published in *The Humanist* magazine[10]:

I am convinced that the battle for humankind's future must be waged and won in the public school classroom by teachers that correctly perceive their role as proselytizers of a new faith: a religion of humanity that recognizes and respects the spark of what theologians call divinity in every human being . . . The classroom must and will become an arena of conflict between the old and new—the rotting corpse of Christianity, together with all its adjacent evils and misery, and the new faith of humanism, resplendent with the promise of a world in which the never-realized Christian ideal of 'love thy neighbor' will finally be achieved.

The modern controversy over the teaching of Darwinism in public schools, while posed as a battle between science and religion, is in fact a continuation of the same cosmic metaphysical battle that probably started at the beginning of time. Today, naturalism—or *humanism* as Dunphy calls it—wrapped in the garb of science, has seized sufficient institutional power in public education that it now controls the content of curricula in the biological sciences, ensuring indoctrination of present and future generations in the fundamental precept of naturalism, the natural origins of mankind.

In the last 100 years, naturalism has ascended to the position of metaphysical ruler over the natural sciences. Why should anyone believe that the scientific priesthood today is going to actually allow, much less ensure, objective instruction in evolutionary biology?

⊷   ⊷   ⊷

Science gave us the potential for great blessings as well as unspeakable horrors, and in the twentieth century we got a healthy dose of both. Science, like any major cultural institution, requires a moral philosophy for its guide. Modern science was developed by Christian natural philosophers in Christian nations. Therefore it was only natural that it was passed down to us within a Christian framework of moral philosophy—a philosophy that generally held that science was to be used for the glory of God and the good of mankind.

Today, however, the scientific establishment rejects any notion of a role for God in our existence and triumphantly rejects His moral law and any other claim He may have on mankind. As we shuffle on down the road to the coming Darwinian utopia in which we will finally be set free from the rule of God, it will be necessary to address the question of morality, especially since it is accepted that the day will come in this new world where Christianity will no longer be a moral and cultural force.

Where will we go to find a reliable basis for establishing this new system of personal morality, social ethics, and political principle? What form of government will we adopt—assuming that we any longer have a choice in the matter? Within this utopian system, we must turn to *science and reason* for the answers where it is taken as a given that the only plausible scientific foundation within such a system for personal and public morality will be some sort of a Darwinian natural selection–based system of morality and ethics.

Such discussions concerning morality and ethics have been carried on for decades among elite intellectuals, and to get a sense of how this discussion is proceeding, you are referred to http://thesciencenetwork. org,[11] which documents meetings carried on at the Salk Institute, La Jolla, California, in 2006, attended by an impressive array of world-renowned scientists such as Richard Dawkins, Paul Davies, Steven Weinberg, Lawrence Krauss, Sam Harris, Michael Shermer, Neil DeGrasse Tyson, Francisco Ayala, and others. This series of meetings was introduced with the following narrative:

Just 40 years after a famous TIME magazine cover asked
"Is God Dead?" the answer appears to be a resounding "No!"

According to a survey by the Pew Forum on Religion & Public Life in a recent issue of Foreign Policy magazine, "God is Winning." Religions are increasingly a geopolitical force to be reckoned with. Fundamentalist movements—some violent in the extreme—are growing. Science and religion are at odds in the classrooms and courtrooms. And a return to religious values is widely touted as an antidote to the alleged decline in public morality. After two centuries, could this be twilight for the Enlightenment project and the beginning of a new age of unreason? Will faith and dogma trump rational inquiry, or will it be possible to reconcile religious and scientific worldviews? Can evolutionary biology, anthropology and neuroscience help us to better understand how we construct beliefs, and experience empathy, fear and awe? Can science help us create a new rational narrative as poetic and powerful as those that have traditionally sustained societies? Can we treat religion as a natural phenomenon? Can we be good without God? And if not God, then what?

This is a critical moment in the human situation, and The Science Network in association with the Crick-Jacobs Center brought together an extraordinary group of scientists and philosophers to explore answers to these questions. The conversation took place at the Salk Institute, La Jolla, CA from November 5–7, 2006.

As you watch these videos you may find that the subtitle, "Beyond Belief," is appropriate. But do not let their lack of believability lead you to think that they are not serious. These people need to be watched.

෫ ෫ ෫

If Darwinism fails as a theory of the origins of mankind, then, as a matter of principle, it cannot possibly provide a sound basis for a moral philosophy and a code of ethics to live by, much less a political system to be governed by. Darwinism does not and cannot account for the presence of mankind on this earth or in this universe. If Darwinism does not account for the origins of man, it cannot account for the true nature of man, and neither can it serve the best interest of man. So, if you see someone coming at you with a moral philosophy, a code of ethics, and a political philosophy based on Darwinism, you must resist with all your strength. It is far more noble to die resisting such an evil than it is to live under it. The only reason our culture is not completely dominated by such evil today is because yesterday many were willing to stand up against that evil, even to the death.

The social experiments of the twentieth century demonstrated conclusively that Darwinism fits in very well with tyrannical systems which naturally suppress ideas about personal rights and liberties among the people. A people who do not want to live in bondage should think very carefully about submitting to an ethical, moral, or political system based on naturalistic Darwinian principles. But if you like that sort of thing or perhaps you think you might like to be the ruler over such a political system, then Darwinism is for you.

CHAPTER 6

NATURAL CAUSES

Restating its basic tenants, methodological naturalism assumes as a working hypothesis that all phenomena observed in the material world are explainable in terms of natural causes alone without reference to the supernatural. Supernatural causes cannot be considered as part of science because science is based on *observation*, and the supernatural cannot be observed. Methodological naturalism functions as a guiding principle that merely imposes practical limitations on scientific methodology—the search for natural causes—and is not meant to be a metaphysical statement about the existence or nonexistence of God.

On the face of it, this appears to be saying that the universe is self-contained and self-regulating and runs on its own, independent of any outside help from God. In this view of the world, everything is explainable within the *mechanical philosophy*. This conclusion is based on more than just brute assumption. This is what is observed, and this is the way the universe actually seems to work. It says nothing however about the origins of the universe or how our earth acquired its properties or how it became a home for a staggering array of living things, including mankind. The existence or nonexistence of a Creator simply does not matter, at least in regard to—this is really important—the *narrow question* concerning those things of interest to the natural sciences, that is, the operation and behavior of the natural world and the laws, principles, and mechanisms that account for that operation and behavior.

The *broader questions* are ontological in nature and are concerned with the ultimate *cause* of the universe, the nature of the universe, and how it acquired its distinctive properties. Methodological naturalism has authority only over the *narrow question* regarding the operation and behavior of the natural world, not its nature or its origins.

But what about life? Can life be viewed as part of the natural world in the same way that matter and the laws of physics are part of the natural world? Can an explanation for the origins of life be found within the constraints of the mechanical philosophy? Did life simply arise as an accident of nature right here on earth?

There is fervent belief that it did, and today this belief dominates the biological sciences. There is this enduring hope, this confidence, that in time scientific knowledge will advance to the point that a naturalistic explanation for how life arose from prebiotic chemicals will be found. Once that fact is established, the place of naturalism in human affairs will finally be firmly established. For the moment, however, origins-of-life research is not all that exciting.

While science cannot presently account for the origins of life, if it turns out that in fact the origins of life can be explained within the mechanical philosophy, then methodological naturalism could legitimately be imposed on scientific studies of the *living* world as well as the *nonliving* world. To put it another way, methodological naturalism has jurisdiction over the living world as well as the nonliving world but *only* if the origins of life and of mankind can be explained within the province of the mechanical philosophy. *This is the central question that underlies the debate over evolutionary biology.* The scientific establishment assures us that it *can* and *does*. The evidence says it *can't* and *doesn't*.

Darwinism faces a dilemma, and this dilemma serves to illustrate why origins-of-life research is not only important to naturalism, it is potentially dangerous to naturalism. What if science never comes up with a credible naturalistic account for the origins of life? It hasn't so far, and there is little prospect that it ever will. Given that the origins of life are unknown scientifically, let us suppose for the moment that there really was substantive evidence supporting neo-Darwinism and that it was no longer controversial among scientists, even among those who believe in creation. Even if we accept that the process of the evolutionary diversification of life might be reliably explained according to Darwin's unguided natural processes, this does not mean that it is *inherently* naturalistic in a metaphysical sense. Evolutionary biologists may argue that the emergence of mankind occurred through natural processes and use that argument in support of a naturalistic view of the world, but ultimately the ontological test concerns the nature of *ultimate* origins: naturalistic or teleological.

> *If the origin of life is teleological in nature, then neo-Darwinism (even if it was true) would be teleological in nature.*

If you are a committed evolutionist and you have no sound naturalistic account of the origins of life—or for that matter, consciousness and mind—you are left with neither a *scientific* nor a *metaphysical* foundation for your theory. There is little prospect that these obstacles will ever be overcome.

Today, in the biological sciences, if we follow the evidence, it leads to a teleological view of nature. The purpose of methodological naturalism is to prevent students from being introduced to evidence, or an argument based on that evidence, that might support that view. Given evidence for design, can methodological naturalism rightly rule out such evidence just because of its teleological implications? No, at least not legitimately, not scientifically. Methodological naturalism has no legitimate jurisdiction over the evidence itself, or an objective interpretation of that evidence which gives no consideration to whatever metaphysical implications might arise from that interpretation. It has authority only over scientific methodology, a methodology, incidentally, that requires that we follow the evidence wherever it may lead in spite of metaphysical implications. If methodological naturalism is used to violate this constraint, then it is functioning as *metaphysical* naturalism, which in effect is a naturalistic religion.

Methodological naturalism deliberately violates that constraint because that is what it has been charged to do. That is its reason for existence, its mission! That covert mission is institutionalized in educational policy and deeply embedded in the Next Generation Science Standards.

❧ ❧ ❧

Our examination of the meaning and effect of methodological naturalism in the natural sciences depends on what we mean by *natural causes*.[1] Does it simply mean "not supernatural"? Can we define its meaning within the context of the natural world as we understand it without reference to the supernatural? The latter surely would be more appropriate because we would like a scientific methodology that is metaphysically neutral and does not depend on metaphysical assumptions.

But this is impossible because, as we saw in chapter 3, the natural sciences were founded on metaphysical presuppositions, such as the existence of a created order, an objective reality, and the rationality and intelligibility of the natural world. What we are expected to do is set aside these presuppositions, adopt the view of the *naturalists*, and establish the meaning of natural causes within the naturalistic framework of understanding.

But before we proceed, there are a couple of matters that must be addressed. The first is an *apology*. The second has to do with *linguistics*.

First the *apology*. My repeated reference to (1) "physical laws, regulating principles and associated mechanisms"; (2) the "operation and behavior of the natural world"; and (3) "observations, experimentation, and inductive reasoning" in this discussion will get tiresome, and I apologize for that, but it is necessary to a single purpose: that of a clear statement and unambiguous understanding of what is meant by *natural causes* and such understanding can only be expressed in these terms. Please bear with me.

Now for the *linguistics*. Because of the different meanings and usages of words like *nature* and *natura*l, discussions of natural causes can easily become muddled in their meaning. One meaning of *natural* is in reference to the material or physical world in its entirety which includes both the nonliving and living worlds. This is the meaning intended when we refer to *natural causes*. A second meaning is in reference to the essential properties of a *thing* (the *nature* of a thing) or its ontology, what makes a *thing* what it is in its essence. A third meaning is in reference to the supernatural where *natural* implicitly means *not supernatural*. Intellectual integrity requires at least a minimum level of care in making sure the intended meaning of slippery words are made clear. In most cases *context* provides a reliable guide, but enough context should be provided to actually convey that guidance.

Now for the meaning of *natural causes*.

The cause-and-effect interactions that continually take place throughout the universe, from the level of fundamental particles to atomic and molecular structure to galaxies and galactic structure, are virtually without number. Because the *causes* in these systems of cause-and-effect chains are themselves directly observable, they are by definition *natural causes*. Much of physics is devoted to explaining the behavior and interactions of systems such as these.

We make a distinction, however, between these systems of directly observable cause-and-effect interactions and the *underlying* physical laws, regulating principles, and associated mechanisms that govern those systems and their interactions. It is these *underlying* physical laws, regulating principles, and associated mechanisms that are the concern of science at its most basic level, and it is at this level that we seek an understanding of what is meant by *natural causes*.

We routinely refer to these underlying *physical laws, regulating principles, and associated mechanisms* as *causes* but it only takes a cursory examination to show that physical laws and regulating principles alone *cause* nothing. It is *mechanisms* that are the causes of physical phenom-enon, and it is an understanding of those mechanisms in terms of the phys-ical laws and regulating principles that govern their actions that constitute scientific explanation at this most fundamental level of physics. As for the makeup of the mechanics of these causal mechanisms, it entails the existence of potential fields (like gravity) and the forces and motions that arise from those fields. The spatial and temporal manifestation of these fields in nuclear, atomic, and molecular structure and their interactions in space–time are what account for the properties of matter and the behavior of the natural world.

No one questions the ability of these natural law-based systems to account for the mechanics of the nonliving world. But are they sufficient in and of themselves to account for life, consciousness, and mind and all that mind entails, such as things like intelligence and reason, and the *higher things* of mind like purpose and meaning, the sense of good and evil, love, joy, wonder, truth, and faith? And what about free will? Is free will just an illusion? The naturalistic view holds that *mind* and all that mind entails is a *natural* phenomenon since it "of necessity" emerged from matter and as such is simply a result of our evolutionary past. The theistic view (specifically the Judeo-Christian view) of the world would say that things, such as life, consciousness, and mind were preexistent in the mind of the Creator and are an intrinsic part of our *being* because we are created in the image of God.

Rene Descartes described mankind in terms of a mind-body dualism leading to the idiom "the ghost within the machine." The body is made of material, nonliving substance that has (so far as we know) no innate capacity for consciousness, self-awareness, sensation, and reasoning. Mind, which is immaterial, not only *inhabits* the body, it *possesses* the body giving it consciousness and life. When I (a "ghost") drive my pickup (a machine), together we are an integrated Cartesian mind-body system. Machines only come "alive" when they are inhabited and possessed by mind.

The reductionist rejects Descartes mind-body dualism and assumes that there is nothing in the natural world, including living organisms, that can be equated to anything *more* or anything *different* than the sum of its parts. Whatever the material constituents may be that make up a living organism, it is a given that those constituents were present in the material world before that organism was formed. Nothing new or different has been added, and nothing has been taken away. Over time, complexity of arrangement developed, and from this complexity, life emerged. Reductionists know that this is true because it *has* to be true. The alternative is unthinkable.

᪥   ᪥   ᪥

While it may be true that *observability* and *testability* are the crucial activities that underlie scientific methodology, and while it may be true that science cannot observe supernatural causes, it is also true that there are essential elements of nature, which themselves are nonmaterial and therefore cannot be observed—for instance, things like physical laws, principles, and mechanisms. These physical laws, principles, and mechanisms not only exist as abstract conceptions of the mind, they are also *present* and *operative* in the natural world. They do not just *appear* to be laws, principles, and mechanisms. They *are* laws, principles, and mechanisms, and they are treated as *real parts* of the natural world because they are operative within the natural world. We know they are operative within the natural world because we observe their effects.

Further, the behavior of these laws, principles, and mechanisms are rigorously described in abstract mathematical terms where the parameters and functional form of the mathematical equations—which themselves are nonphysical and can only be "observed" and understood within the realm of mind—are deeply embedded in physical reality. But if we cannot observe the physical laws, principles, and mechanisms that govern the operation and behavior of the natural world, how then is science to proceed?

Causal mechanisms and their governing physical laws and principles are *inferred* through observation of the *effects* they produce in the natural world. We don't *see* the laws, principles, and mechanisms themselves; we *see* their *effects*, and from those effects we infer their existence and their properties. Neither do we observe the equations of physics in nature. Rather, we observe relationships between physical parameters that can be described mathematically, and from those observations, we infer the mathematical form of these equations and the values of any physical constants they may contain, all determined through observations in which physical parameters are quantified through carefully designed experiments.

> *Science observes effects and infers causes.*

Newton's law of gravity is an example of *physical law*. Conservation of energy is an example of a *regulating principle*, and the attractive forces resulting from the gravitational fields associated with massive bodies (referring to mass as an intrinsic property of matter) is a *mechanism* that acts in the natural world. The motions of a planetary system and the motion of a pendulum are examples of observed *effects*, describable through Newtonian mechanics. Through this kind of reasoning, we conclude that all *observable* natural phenomena are the result of natural causes. They are *natural* causes because they are present and operative in the *natural* world, and we know that they are present and operative in the *natural* world because we observe their *effects*.

So, when it comes to the *fundamental* laws, principles and mechanisms of nature, it is in this sense and only this sense that we can say that science *observes* causes in nature.

ଶ   ଶ   ଶ

So, what do we mean by *natural causes*? Let us proceed by agreeing that we should assume a *model* of the natural world (the nonliving world) as *a closed, self-contained system of cause-and-effect relationships that operates moment-by-moment on its own, according to fundamental laws, regulating principles, and associated mechanisms (causal mechanisms) without intervention by any transcendent intelligent agency.* Basically, this is the *mechanical philosophy,* the view that eventually emerged among Christian natural philosophers of the seventeenth and eighteenth centuries in their efforts to develop an understanding of the relationship between God and His creation. It is also consistent with both the *greater-God* hypothesis and the *no-God* hypothesis.

This closed self-contained model of the natural world also provides the framework for scientific methodology. Accordingly, a definition of *natural causes* that "fits" that framework and does not entail any hidden metaphysical assumptions will be metaphysically neutral. And here it is:

> ***Natural causes are those causes that are present and operative in the natural world that produce observable effects that can be subjected to scientific study.***

This *rational* definition of *natural causes* makes no judgment one way or the other about the supernatural. If a cause is present and operative in the natural world such that it produces observable effects that can be subjected to scientific investigation, then scientifically it may rightly be called a *natural cause*. Natural causes so defined are restricted to the *narrow question* concerning the operation and behavior of the natural world and make no reference to the b*roader ontological questions* concerning the *ultimate cause of* or *nature of* the natural world. The *ultimate cause* behind these *natural causes*, whether it be material or supernatural, is forever beyond the reach of the natural sciences and consequently forever *off limits* to methodological naturalism. *The concern of methodological naturalism is scientific methodology within the natural sciences, not the ontological properties of the natural world.*

Scientifically, the understanding of natural causes framed above is consistent both with the views held by seventeenth- and eighteenth-century natural philosophers who believed in God, and present-day naturalistic scientists who do not. We can accept that in principle methodological naturalism is a metaphysically neutral concept but only if (1) it is restricted to the *narrow* questions concerning those things which can be observed by the methods of science, and (2) ontological implications inferred from scientific observations have no bearing on the scientific merits of an investigation.

∽ ∽ ∽

In such a discussion, it is inevitable that at some point the so-called "God-of-the-gaps" theory will arise. The idea here is that whenever anyone points out some aspect of the natural world that cannot be explained through the laws of physics and chance alone and notes that it exhibits features that are normally attributed to design, then such reference to design is met with the accusation that this is simply a claim that *God did it*. Such appeals to the supernatural, they say, are "science stoppers" that lead to dead ends and prevent further legitimate scientific investigation. For anyone who actually thinks this way, this may be true, but the same could be said for "Darwin-of-the-gaps" and its claim that "chance did it" which, incidentally, is the ruling paradigm in the biological sciences.

But "God-of-the-gaps" is little more than a cheap rhetorical device and a means for dodging serious discussion while trying to leave the impression that you have said something significant. No serious scientist ever proposed that "God did it" counts as a scientific explanation of any natural phenomena. However, many serious scientists claim as a meta-physical necessity that *chance* and *time* are able to do those things that only God could do if there was a God to do them. Is evolutionary biology just a naturalistic version of "God did it"? We must admit that the abilities claimed for chance processes seem miraculous.

There never was and there never has been a philosophy of science that would permit such a superficial and transparent ploy as that represented by the "God-of-the-gaps" argument. It is interesting to note, however, that most of the natural philosophers of the sixteenth, seventeenth, and eighteenth centuries would agree that from the larger picture of creation, God in fact did *do it* and in fact *did it all*. That was a given. The question of concern to the natural sciences at that time was *how* He did it and how mankind could use that understanding for the glory of God and the good of mankind.

But gap theory aside, there is an important question here. Given the design-like features observed in biology that everyone admits to, how should science treat such observations?

It turns out that in the face of evidence that simply could not be ignored, scientists in the biological sciences set aside their inherent antagonism toward design principles and decided to get on with the science. They followed the evidence. That particular area of research is called *systems biology*.

The field of systems biology,[2] which emerged in the biological sciences during the 1990s was founded, not on chance-based Darwinian principles, but on design and information principles taken from the well-established fields of *systems engineering* and *information theory*. Systems biologists may give a metaphysical nod to Darwin but in their science, they follow design principles, and no one calls them heretics because they do.

Scientists follow the evidence. Priests defend the dogma. It is not difficult to tell the difference. In the 1990s biologists followed overwhelming evidence of design in cellular biology, and it led them to an entirely new discipline within the biological sciences. Today, systems biology is where the cutting edge research within the biological sciences is performed. This new discipline also constitutes a de facto admission by science that *actual* design is present in nature (whatever its cause may be).

⋰ ⋰ ⋰

While we may accept that God does not intervene in the natural world concerning its day-to-day workings, that does not mean that He is not active in creation in ways that are invisible to the methods of science. Perhaps He goes about the divine business of forgiving sinners, weeping over us, redeeming souls, giving hope, seeing us through the hard times, pointing the way in confusing times, listening to prayers, watching over His children, calling us home when our time comes, and maybe every now and then a miracle of healing, either of the body or of the soul, all completely under the radar of science. But as for the natural world, He allows it to run on its own. No tinkering. No tweaking. No orbits to adjust. Good things happen, bad things happen, but the natural world just keeps on rollin' along.

This does not mean that there were no *intervention* events in the remote past associated with the *process* of creation, the special ordering of the natural world, and the creation of life. Who really knows? We can only speculate about such things. And who among mortal men is to say what the Almighty may or may not do regarding His creation?

Further, there are around seven billion *intelligent agents* living today in a highly localized region of the cosmos (on the surface of the earth), and throughout their lives, while they may not tinker with the constants of nature or sling comets around the solar system or set off supernovas, they do influence the natural world. They come into this world, they live their lives, they build things (or destroy things depending on their disposition), they rearrange a few things, they leave their footprints and fingerprints here and there, they die, and then they return to nature in complete harmony with the laws of physics and chemistry. No laws of physics and chemistry have been broken or interrupted. The natural order is still intact. The cosmic machinery continues to grind along. Yet, during their lives, they change the world in ways that are beyond the reach of the laws of physics and chance alone. Because of each life, the universe is forever changed, maybe only in a relatively microscopic sense, but nevertheless, it is changed.

If human intelligent agents can intervene in the natural world and create technological wonders in harmony with the natural order, on what grounds might we argue that the Creator cannot do the same?

∽ ∽ ∽

Reason tells us that whatever the ultimate cause of the universe may have been, the universe itself will reflect the nature of that cause. The material world will testify as to its true ontological nature: meaningful or meaningless, purposeful or purposeless, teleological or materialistic. It is what it is. We cannot make it to be other than that which it is no matter how abhorrent the metaphysical message of the universe might be to some. The only thing that is certain is that there *is* such a *message*, science can "read" that *message*, and not everyone is going to like what it has to say. That is just the way it is.

Here at the boundary between the material and nonmaterial realms, between the natural and the supernatural, between the universe and its cause, between life and its cause, and between mankind and his cause, science encounters its limitation. That limitation is defined by what science can and cannot observe. This boundary is like a *veil*, a veil beyond which science cannot see. We may make inferences about that which lies beyond that veil based on what we observe on this side of the veil, but scientifically, that is as far as we can go. Methodological naturalism has legitimate jurisdiction over the methods of science that may lead to a metaphysical inference but *it has no jurisdiction over the inference itself.*

Whatever the *cause* of the universe might be, that cause has left its fingerprints all over creation. They may be naturalistic fingerprints or they may be divine fingerprints, but there *are* fingerprints. We can both *sense* their presence in the general attributes of the natural world as in Psalms 19:1 "the heavens declare the glory of God" or if you prefer, in the words of Carl Sagan "The Cosmos is all that is or was or ever will be," and we can "see" them through the "eyes" of the natural sciences.

> *The search for fingerprints and attempts to discern that to which they point is a perfectly legitimate activity of science.*

While the search for fingerprints may be a perfectly legitimate activity of science, the ontological inference that may arise from such activity is off-limits to methodological naturalism. Thus, if a finding of science reveals design-like features in biology and the scientific methodology employed in the investigation that led to that discovery was subject to the constraints of methodological naturalism, that is, the normative methods of observation, experimentation and inductive reasoning, then methodological naturalism has nothing more to say. Otherwise it is metaphysical naturalism. In fact, the appeal by scientists to *chance* in attempting to explain so many things about the living world is a *metaphysical* necessity, not a *methodological* necessity.

Our expectation is that there will be a deep coherence between what we observe on this side of the veil and that which lies beyond. If nothing lies beyond, then there will be nothing on this side to give evidence that there is *something* "over there." We can never directly confirm this coherence scientifically, but we are nevertheless confident that it is true. How can we make such an assertion? Because this is what we would expect in a rational universe. Following C. S. Lewis' argument, if this was *not* a rational universe, we would not know it was *not* a rational universe, and it is quite certain that we would not even be here to carry on such a discussion. In short, it just makes sense—overwhelming sense. It is self-evident. We are here!

≈ ≈ ≈

By all appearances and in light of the lack of any plausible naturalistic account of life, mind, and consciousness, it is reasonable to suppose that they had their origins in a preexisting transcendent mind with the capacity to create our present material universe with its laws, principles, and mechanisms, all according to some master plan and purpose. Present evidence of both physical law and design in nature testify to the past work of intelligent agency. As will be shown in the chapters to follow, for those who are willing to approach this subject with an open mind, there is no longer any doubt that the cosmos has a message for us, a message that is good news for mankind.

CHAPTER 7

THE INTELLIGENT DESIGN MOVEMENT

I n 1959, scientists from around the world gathered at the University of Chicago to celebrate the Darwinian centennial. It was an extraordinary event heralding not only the 100th anniversary of Charles Darwin's publication *On the Origin of Species,* but also the dawning of a new age in the history of the earth and of human affairs.

Sir Julian Huxley, grandson of Thomas Huxley captured the excitement and historic nature of the occasion in his address which opened the symposium.[1]

> Future historians will perhaps take this Centennial Week
> as epitomizing an important critical period in the history of
> this earth of ours—the period when the process of evolution,
> in the person of inquiring man, began to be truly conscious of
> itself. . . . This is one of the first public occasions on which it
> has been frankly faced that all aspects of reality are subject to
> evolution, from atoms and stars to fish and flowers, from fish

and flowers to human societies and values—indeed, that all reality is a single process of evolution. . . . In the evolutionary pattern of thought there is no longer either need or room for the supernatural. The earth was not created, it evolved. So did all the animals and plants that inhabit it, including our human selves, mind and soul as well as brain and body. So did religion . . .

There was good reason to celebrate—or so everyone thought at the time. In the preceding decades Mendelian genetics had been wed to Darwin's natural selection to produce an updated version of Darwinism called "the Modern Synthesis" or "neo-Darwinism." Biologists now understood the basic mechanisms of heredity, a crucial aspect of Darwinism that had been missing ever since publication of Darwin's book. In 1953, Watson and Crick discovered the double-helical structure of DNA opening the door that would lead to an understanding of the very secrets of life. Also in 1953, Miller and Urey published their work[2] on production of amino acids from inorganic precursors by subjecting them to atmospheric conditions that supposedly existed early in the history of the earth, hinting at a mechanism that may have led to the emergence of life through natural processes alone.

In the years following the Chicago centennial celebration, biologists cracked the genetic code, and by the late 1960s, evolutionary biologists were sure that they were now on a path that would lead to a completely naturalistic understanding of the origin of life. But trouble was brewing, big trouble.

෴ ෴ ෴

These advances in molecular biology—while rejuvenating Darwinism by identifying the mechanisms for heredity and the preservation of genetic change—also revealed the staggering complexities of the biological cell and the mechanisms of the life processes. An important question was taking shape: How might neo-Darwinism be reconciled with this new understanding of biology?

These developments in biology did not escape the notice of world-class mathematician Murray Eden at MIT. In the early 1960s, Eden and his colleagues, along with Marcel Schutzenberger in France, began to model the Darwinian mutation-selection mechanism and were immediately impressed by the consistently negative results produced by their computer simulations. These findings eventually became known to some evolutionary biologists and in good scientific tradition, a meeting was planned to review those results.

In 1966, a historic meeting of evolutionary biologists and mathematicians called the Wistar Symposium,[3] was held at the Wistar Institute in Philadelphia to review the findings of the mathematicians. We get a hint as to the mood of that meeting from what happened when Schutzenberger—well known for his fearless and outspoken manner—stood up and announced that "*there is a considerable gap in the neo-Darwinian theory of evolution, and we believe that this gap cannot be bridged within the current conception of biology.*" Darwinist C. H. Waddington set a historic precedent concerning how Darwinists should respond to such challenges when he replied "*Your argument is simply that life must have come about by special creation.*" The mathematicians immediately responded in unison with a loud "No!" and it went downhill from there. Eden, Schutzenberger, and the mathematicians knew that their computer simulations had exposed a fundamental problem with Darwinism, and they would not allow the biologists to dodge this matter by attempting to turn it into it a dispute between science and religion.

The mathematicians stood firm in their findings, while the biologists were convinced that the mathematicians had something wrong with their equations but had no idea what it might be. After all, they *knew* that the eye had evolved; thus, the computer simulations performed by the mathematicians *had* to be wrong. That was the only possible conclusion. Ernst Mayr consoled the biologists by pointing out that "we can take comfort in the knowledge that evolution is true," but that was the only comfort they took from Wistar.

Wistar resulted in a stalemate and was followed in 1969 by another symposium[4] at Alpbach, Austria, this time attended only by evolutionary biologists. The mathematicians were not invited. Even without the mathematicians, Alpbach resulted in intense disagreements, verbal brawls, hard feelings, and again—stalemate. But there was one thing that all agreed upon. Neo-Darwinism had some problems, and it was not clear how they were going to be resolved.

Troubles continued throughout the 1970s. Pierre Grasse[5], French zoologist and one of the most respected biologists on the planet, dropped a bombshell in the biological world when in his book *The Evolution of Life* (1977 English edition), he said:

> It is possible that in this domain, biology—impotent— yields the floor to metaphysics.

Two additional meetings were held in the spirit of Wistar and Alpbach, the first in October 1980 at Chicago's Field Museum of Natural History and the second the following year in New York at the American Museum of Natural History. The purpose of the Chicago meeting was to address a fundamental question: Can the mechanisms underlying microevolution be extrapolated to explain macroevolution? The conclusion of the majority was *no* and that there was no evidence of evolution, the fossil record did not support evolution, and there was no way to demonstrate that it *is occurring* or *has ever occurred*. While it was decided that no records of the proceedings would be kept (to avoid giving ammunition to the creationists), this event, nevertheless, was widely reported such as in *Newsweek*,[6] the journal *Science*,[7] and books such as *The Neck of the Giraffe*[8] and *Great Evolutionary Mystery*.[9] The results of the New York meeting further confirmed the disarray of evolutionary theory. It was in this meeting that Colin Patterson, senior paleontologist at the British Museum of Natural History, made his famous statement "all my life I had been duped into taking evolution as revealed truth." In 1981, Patterson traveled from conference to conference,[10] asking the same embarrassing question and always getting the same embarrassing answer: silence. His question was:

Can you tell me one thing about evolution that is true—any one thing at all?

It was becoming apparent to the evolutionists that what they had was a theory of biology that they claimed was the cause of *everything* in biology, but it seemed to explain *nothing* about biology.

101

Astronomer and hard-core atheist Fred Hoyle[11] and mathematician Chandra Wickramasing wrote in their book *Evolution from Space* published in 1984:

> The speculations of "The Origin of Species" turned out to be wrong . . . It is ironic that the scientific facts throw Darwin out but leave William Paley, a figure of fun to the scientific world for more than a century, still in the tournament with a chance of being the ultimate winner.

Harvard paleontologist Stephen Jay Gould expressed his understanding of the problems with the fossil record as unambiguously revealed by the greatest paleontological discovery of the twentieth century, the Burgess Shale. In his essay in *Natural History*[12] he wrote:

> The extreme rarity of transitional forms in the fossil record persists as the trade secret of paleontology. The evolutionary trees that adorn our textbooks have data only at the tips and nodes of their branches . . . in any local area, a species does not arise gradually by the gradual transformation of its ancestors; it appears all at once and fully formed.

This statement by Gould would plague evolutionists for decades to come—and still does.

Paleontologist Niles Eldredge and paleoanthropologist Ian Tattersall state unequivocally in their book The Myths of Human Evolution[13] published in 1984 . . .

> The record is there, and the record speaks for tremendous anatomical conservatism. Change in the manner Darwin expected is just not found in the fossil record. . . . species are stable and remain discrete, in time as well as space.

By the 1980s, all of the relevant data made it clear that natural selection was unable to account for the complexities of the biological cell (not to mention macro-evolutionary change) and neither could the actual history of life as recorded in the fossil record be reconciled to Darwin's common ancestry descent description of that history. In effect, by the mid-1980s, the best scientific evidence suggested that *neo-Darwinism was little more than a speculative hypothesis about a theory that was unable to explain something that never happened.* This statement of course is nonsense, but that is the point.

To make matters worse, things were not going well in the area of origin-of-life research. By this time, scientists had concluded that the gases that actually existed in the early atmosphere at the first appearance of life were different in important ways from those used by Miller and Urey in 1953 in their original origin-of-life experiments. Miller and Urey had simply assumed a composition similar to that of interstellar gases with no oxygen because even a small amount of oxygen in the atmosphere would interfere with the formation of organic molecules or would destroy any that did form. It was becoming increasingly evident through the study of ancient geologic formations and theoretical analysis of photo dissociation rates of water that it is very likely that the earth's atmosphere has contained some significant level of free molecular oxygen over 99 percent of its life. In spite of these difficulties, the Miller-Urey experiments have continued to appear in high school and university textbooks to advance the idea of the origin of first life through natural processes.

In 1984, Charles Thaxton, Walter Bradley, and Roger Olsen published their book *The Mystery of Life's Origins*.[14] In this book, a highly technical work, the authors rigorously applied the fundamental laws of thermodynamics to the problem of chemical evolution. The power of their methodology was that the outcome was based on elementary thermodynamic analysis as opposed to imaginary chemical pathways that supposedly connected living organisms with nonliving chemicals. Their Second Law analysis demonstrated the extreme implausibility of chemical evolution, and they concluded that the "prebiotic soup" was a myth. There never was such a "soup."

This work by Thaxton, Bradley, and Olsen stands even today, thirty years later, without refutation. No one has even tried, or at least not in any major peer-reviewed published work.

In 1986 Robert Shapiro (no creationist he) in his commentary on this research wrote in his book, *Origins: A skeptic's guide to the creation of life on Earth*[15]:

> We shall see that the adherents of the best-known theory have not responded to increasing adverse evidence by questioning the validity of their beliefs, in the best scientific tradition; rather, they have chosen to hold it as a truth beyond

question, thereby enshrining it as mythology. In response, many alternative explanations have introduced even greater elements of mythology, until finally science has been abandoned entirely in substance, though retained in name.

Note that Shapiro has moved on from arguing about the evidence for biological evolution (which is essentially nonexistent), to scathing criticism of the shameful behavior of evolutionary scientists themselves.

In spite of all of these difficulties, in Darwinism's darkest hour, something inexplicable happened. Out of the rubble of the scientific critique of the 1960s, '70s and '80s, at its historically weakest moment, Darwinism was "raised from the dead" and its position in public education secured. Darwinism was saved, not through scientific discovery, but by the United States Supreme Court.

In 1987, the Supreme Court, in its findings in *Edwards v. Aguillard* (which overthrew a Louisiana statute requiring balanced treatment in teaching both evolution and creation in public schools) declared that the Louisiana statute was a sham and therefore had a religious rather than a secular purpose. Consequently, it was unconstitutional. The net effect of the *Edwards* decision was to establish Darwinism, by default, as the official creation story of America. On that day, a secular religion was established in America where the public schools (unknowingly) became in effect its educational and missionary arms, the universities (knowingly) became its seminaries, and the natural history museums (knowingly) became its cathedrals. And the best part? It was all paid for through taxes collected by the government from the parents of the children being targeted for this indoctrination. *Edwards v. Aguillard* will be reviewed in detail in chapter 13.

❧   ❧   ❧

Earlier, in chapter 3, we made the point that in the natural sciences the merits of the evidence are central to establishing the credibility of a theory of science, and surely this should be a concern of the highest order in science education in public schools, especially for a controversial theory such as Darwinism with its disturbing religious implications. In consideration of that concern and in light of the exalted legal status given to evolution by the Supreme Court, it is appropriate at this point to contrast the legal status of evolution with its scientific status based on the merits of the evidence. This contrast serves to point out the utter absurdity of the situation that was created in our public schools and has reigned for thirty years.

The fossil record is crucial to establishing the validity of any theory of biological origins that claims to give a scientific account of the history of life because it is the fossils that provide the only *direct physical evidence* regarding that history. The Darwinian crisis of the 1960s, '70s and '80s painted a grim picture of that record, and nothing has changed since then. Without a substantive base of data to fill in the evolutionary narrative, there is no scientific theory of biological evolution. You might get a tentative hypothesis, which is what Darwin provided, but not a full-blown scientific theory. But this is *not* the way it is presented in public education, which censors any evidence or interpretation of evidence that is critical of evolution.

In Darwin's day, it was well known that the fossil record presented serious challenges to his ideas about slow gradual changes and innumerable intermediates. Darwin expressed his hope that further discovery would fill in the missing intermediates. But after more than 150 years of searching the rocks, the situation is worse, not better, and no one believes that in another 150 years that things will change for the better.

The problems with the fossil record, such as the sudden appearance of the great variety of Cambrian animals in the Burgess Shale 435 million years ago followed by progressive major extinction of most of the Burgess animals, does not conform to Darwin's expectation concerning the history of life. Sudden appearance of fully formed species with no apparent ancestors, followed by stasis and extinction are, according to leading paleontologists such as Stephen J. Gould, Niles Eldredge, Colin Patterson, and Robert Carroll, the dominant features of the fossil record. Only by incorporating a superficial selective examination of the data (as is typically done in high school biology textbooks) can the fossil record be reconciled to Darwin's predictions. See Stephen J. Gould, *Wonderful Life*[16]; Douglas Futuyma, *Evolutionary Biology*[17]; and Robert Carroll, *Patterns and Processes of Vertebrate Evolution.*[18]

We take specific quotes from Robert Carroll, professor of paleontology at McGill University, who is probably the foremost living expert in the world on vertebrate paleontology. In 1997 he published his book *Patterns and Processes of Vertebrate Evolution* as part of the prestigious Cambridge Paleobiology Series of scholarly works, which has become a minor classic in its field. In the preface and first chapter of his book, Carroll makes some extraordinary statements about pervasive problems in the history of life constructed from examination of the fossil record compared to the history predicted by Darwin. He even criticizes textbooks (p. 2) for perpetuating this false history. Here is Carroll:

Instead of showing gradual and consistent change through time, the major lineages appear suddenly in the fossil record, already exhibiting many of the features by which their modern representatives are organized. (p. 2)

Few fossils are yet known of plausible intermediates between the invertebrate phyla, and there is no evidence for the gradual evolution of the major features by which the individual phyla or classes are characterized. (p. 4)

Progressive increase in knowledge of the fossil record over the past hundred years emphasizes how wrong Darwin was in extrapolating the pattern of long-term evolution from that observed within populations and species. (p. 8)

This theme is repeated in all fifteen chapters of his book.

High school biology textbooks will usually say little about the fossil record, but what is said will skirt these issues while being careful to leave the impression that macroevolution is supported by the fossil evidence. Whatever may have happened in the past, the history of life revealed in the dominant features of the fossil record does not reflect Darwin's concept of universal common ancestry descent. Darwin, to his credit, was very careful to speak hypothetically about his theory and the last 150 years of research have confirmed that he was right in doing so.

How can public education reconcile what is taught in high school biology classes with these facts as stated by leading paleontologists? The deception perpetrated by public education against the children of America is deliberate, it is explicit, it is endemic, and it is potentially conspiratorial. I understand that there are laws against such a thing, but I guess that when you have the Supreme Court as a partner in the deception, you probably do not have to be concerned.

ఆ   ఆ   ఆ

If Darwinists, as scientists, are truly concerned about the intellectual and academic integrity of their discipline, then they need to start acting like scientists. Give up the bullying, the pretense of authority, the wild speculations, the deceptive arguments, gang-land style tactics, and running off to the courts to fight your battles for you on legal terms since you can't win on scientific terms. If Darwinism is to be accepted as a legitimate scientific theory concerning the history of life, then the validity of that theory can only be established through an objective consideration of the physical evidence. The physical evidence that makes or breaks Darwinism is the *fossil record*. What is the point of trying to explain something that obviously never happened? The fundamental undisputed facts taken from observations of the raw fossil data are that we see an astonishing variety of *different* species with wildly different body plans, anatomical features, reproductive means, and modes of life that appear in the fossil record suddenly and fully formed, they remain the same (stasis) throughout their tenure on earth (which in many cases is for tens of millions of years), and then, in the vast majority of cases, they go extinct. The principal features of the history of life on earth based on what we *observe* is that this history can be reduced to *differences and stasis*, not *continuity and change*—or, to reduce it even more, *creation*, not *evolution*. In either case that's a big problem if you do not believe in God, or you do not *want* to believe in God, or you do not want school children to believe in God.

The validity of Darwinism as presently accepted within the natural sciences is *not* established by an objective consideration of the merits of the physical evidence. It is established by the assumption that the theory is true and that the history of life *must* be interpreted within the framework of the theory, not the physical evidence. This is dogma, not science, and this is what is being taught in our public schools.

In *Biology* by Campbell and Reese,[19] a top-tier high school biology textbook, five chapters are devoted specifically to evolutionary theory. Their presentation is typical of that found in most other high school biology textbooks. However, through a process of literary *infiltration*, they have contrived to embed evolutionary dogma in all of the subtopics of the book and throughout all fifty of its chapters. Thus, the impression is given that evolution is much more than just a theory about the history of life. Rather it is a fundamental mechanism of biology in the same way that gravity is a fundamental mechanism of cosmology. You cannot understand the universe without gravity, and you cannot understand biology without evolution. That is the message. One of these statements is fact. The other is dogma. I think the reader will have little difficulty in figuring out which is which.

The final step in this strategy of *infiltration* is being implemented in the Next Generation Science Standards (NGSS) where, as of this writing, twenty-six states have adopted those standards where evolutionary dogma is systematically and progressively introduced, a bit at a time, at every grade level.

The *infiltration* strategy is a masterful means for securing and holding institutional power, and if naturalistic dogma is to be eradicated from public education, it must be done systematically by eradicating dogma itself and that means dogma of any kind. But what we all have to understand is that eradicating naturalistic dogma does not mean targeting evolution for eradication. Rather, it just requires that evolution be taught objectively as science, not as dogma, permitting an open and objective consideration of the merits of the evidence regardless of whether that evidence supports evolution or not. Objectivity in origins science education is not an option. Theories stand or fall on the merits of the evidence, not their metaphysical implications or their political utility.

In the end, the Darwinian empire will fall. What will bring it down is the evidence treated within a scientific community committed more to the ideals of intellectual, scientific, and academic integrity than to naturalistic dogma. But the scientific priesthood, the science teachers' unions, progressive politicians and judges, and many who hold to a materialistic worldview cannot allow this to happen. There is much more at stake than science. This is about worldview, institutional power, and politics. If the matter was about science alone, it would have been settled long ago.

In the end, from the perspective of evolutionary scientists, the most important feature of any evidence is that it must *not* support the teleological view of nature. Consequently, this "must not" feature of evolutionary biology is deliberately embedded in the teaching of evolutionary biology in public schools. In fact, since the evidence for evolution is so scarce and the evidence for design is so pervasive, of necessity, evolutionary science is more of an *anti-science* that concerns itself with teaching students to accept the *dubious* as *fact* and to reject the *self-evident* as *illusion*. Consequently, curricula in evolutionary biology must be arranged accordingly with the ultimate objective of slow, painless, and insidious eradication of the teleological view from the minds of students, a way of thinking about the natural world that has dominated Western thinking for over 2500 year.

∙๑ ∙๑ ∙๑

So, what does all of this have to do with intelligent design? Darwinists claim that the intelligent design movement was formed by creationists who, after their defeat in *Edwards v. Aguillard,* secretly got together and devised a plan to make an end run around the Supreme Court and sneak creationism, under a different name and with a different "look," back into the public school science classroom. The new name was *intelligent design* and the new "look" was biological design.

This narrative is pure fiction. A review of the actual history of this period reveals that the origins of the intelligent design movement can be traced back to the dissent from Darwinism within mainstream science that took place in the 1960s, '70s and '80s. The explosive increase in knowledge about cellular biology during this period equated to explosive increase in bewildering problems for Darwinism. Then in 1985, Michael Denton, a young biochemist with a PhD from King's College, London University, came out of nowhere, and walked out to center stage in the creation-evolution brawl with the publication of his book *Evolution, a theory in crisis*.[20] In his book, Denton provided a broad and devastating scientific critique of Darwinism while at the same time openly acknowledging evidence of design in nature and the rationale for the teleological view of the world. The release of Denton's book was one of the key events leading to the emergence of the intelligent design movement. Because of the sheer breadth and depth of Denton's work, it could not be dismissed. And it wasn't. It still stands. It was historic. It could very well be remembered as the book that eventually took down Darwinism.

The intelligent design movement took shape in the late 1980s and early 1990s through various conferences and meetings and the formation of the Discovery Institute in Seattle, Washington, which became the intellectual center for the movement. In 1991 Berkley legal scholar Professor Phillip Johnson's book *Darwin on Trial*[21] was released, taking the issues well beyond the borders of science to the general public where it was widely read. Johnson, with masterful skills in logical argumentation, artfully dissected Darwinism to expose not just a flawed scientific theory but an underlying naturalistic philosophy that rules all scientific work. Johnson's book was followed by release in 1996 of Michael Behe's book *Darwin's Black Box*[22] where he introduced the concept of irreducible complexity (and his famous mousetrap) laying down a powerful biochemical challenge to evolution. Johnson and Behe were both inspired to write their books when they read Michael Denton's *Evolution: a theory in crisis*. Then in 1999, mathematician and philosopher William Dembski published his book *Intelligent Design*[23] that put the theory on rigorous mathematical and logical grounds and contributed greatly to defining the movement and giving it its identity.

With the formation of the Discovery Institute and the release of Johnson's, Behe's, and Dembski's book (all inspired by Denton), a vector had been set for the intelligent design (ID) movement. The battle had been engaged on legal, philosophical, and scientific fronts—and there was much more to come. ID was on the move. Grassroots organizations popped up around the country, technical conferences were organized, high quality DVD documentaries on ID were produced (e.g., Illustra Media), and a wave of scholarly books related to ID were published such that today there is a wealth of resources documenting every aspect of this controversy.

Intelligent design as an intellectual movement is here to stay.

∽ ∽ ∽

We have reviewed the history of the intelligent design movement and how it emerged from the dissent within Darwinism. Now we look at ID itself and how its scientific content challenged Darwinism.

As long as science had a superficial understanding of biology, Darwin's theory was plausible. And, while his theory was subjected to considerable criticism, knowledge in the biological sciences at the end of the nineteenth century was insufficient to establish Darwinism as either fact or fiction. Darwin was careful in stating his claims. He never made other than hypothetical arguments and was first to point out the weaknesses in his theory—weaknesses that still persist today. Thus, for ninety years after its introduction to the world, Darwinism, at least in its scientific form, survived on an uneasy diet of ignorance. Then, around mid-twentieth century, spectacular advances in the field of molecular and cellular biology brought new understanding to the complexities of living things. Neo-Darwinism was benefitted greatly by these advances through an understanding of the biological mechanisms of heredity and variation and how those variations were preserved from one generation to the next. But a troubling downside accompanied these advances in understanding.

The complexity of the mechanisms of heredity presented a staggering challenge, not only to Darwin's naturalistic account of the *history* of life, but also to the problem of providing a naturalistic account of the *origins* of life. How could such complex systems emerge through natural unguided processes alone?

Ultimately, the questions of the origins of life and the origins of mankind cannot be separated. Naturalism requires a coherent description of both. Without a convincing account of the origins of life (which obviously came first), whatever account that may be given for the origins of mankind (which came later) will be left without a foundation. When we consider the implausibility of neo-Darwinism in light of our present understanding of biology and the paucity of the physical (fossil) evidence supporting common ancestry descent, this naturalistic account of life is on exceedingly precarious scientific grounds. Yet it is presented to students as undisputed scientific truth.

The challenges to Darwinism that arose during the 1960s and that grew into full-blown dissent in the 1970s and 1980s as previously described introduced a hitherto unknown and completely unsuspected feature of biology. Living things not only had the appearance of having been designed—no one disputed that—but by the end of the twentieth century, it had been established that design *principles* were present and operative deep down in the fundamental machinery of life.

The nature of the discussion over design in biology was fundamentally transformed from that of philosophical speculation about the way things ap*peared,* to scientific investigation as to how things actually *work* based on what was observed. Darwinism could no longer hide under a blanket of ignorance. If Darwinism was to survive as a legitimate scientific theory, it had to give an account of how these complex systems arose through unguided natural processes alone. Otherwise, *there is no theory of organic evolution.*

Darwinism's status in American public education today is secured not by credible scientific evidence and logical argumentation but by cultural power wielded by a materialistic scientific establishment defending their naturalistic dogma, court rulings based on outlandish and deceptive arguments, the myth of a war between science and religion, misguided applications of First Amendment law (to be addressed), and even progressive political philosophy. Powerful nonscientific interests have a vital stake in evolutionary theory.

∽ ∽ ∽

The prevalence and uncontroversial nature of evidence of design in biology is reflected in statements by leading evolutionary biologists and philosophers where their statements reflect our *modern* understanding of molecular and cellular biology rather than the simplistic understanding of Darwin and his contemporaries. They also reflect an unshakable commitment to naturalism.

Richard Dawkins says in his book *The Blind Watchmaker*[24]:

> Biology is the study of complicated things that give the appearance of having been designed for a purpose.

Here is Douglas Futuyma on design from his popular university textbook *Evolutionary Biology*[25]:

Darwin's immeasurably important contribution to science was to show how mechanistic causes could also explain all biological phenomena, despite their apparent evidence of design and purpose

Daniel Dennett, professor of philosophy at Tufts University in 2005 wrote[26] in the *New York Times*:

Natural selection has the power to generate breathtakingly ingenious designs.

Francis Crick, Nobel Laureate, said[27]:

Biologists must constantly keep in mind that what they see was not designed but rather evolved.

Francisco Ayala, in his article "Darwin's Revolution" in *Creative Evolution?* wrote[28]:

The functional design of organisms and their features would therefore seem to argue for the existence of a designer. It was Darwin's greatest accomplishment to show that the directive organization of living beings can be explained as the result of a natural process, natural selection, without a need to resort to a Creator or other external agent.

Here is Ernst Mayr on design from his article in *Scientific American*[29]:

> Design in the natural world was so convincing that it inspired a form of apologetics called natural theology. Natural theology had enormous influence on most of the field work done in biology from 1650 until 1850.

> It is difficult for the modern person to appreciate the unity of science and Christian religion that existed from the Renaissance and far into the eighteenth century. . . . The Christian dogma of creationism and the argument from design coming from natural theology dominated biological thinking for centuries.

∽ ∽ ∽

One hundred years ago, there was little other than speculative Darwinian arguments (known as "just-so stories") and hand waving (an essential element of good story telling) to support the Darwinian view of life. Today, however, molecular and cellular biologists have "raised the hood" on the biological cell to expose its inner workings where they have found molecular systems of staggering complexity, which reflect design-like features that Darwin could not have imagined.

These scientific discoveries have taken the *design argument* to a completely different level. Two-hundred years ago, design and purpose in nature were inferred from the general harmony observed between the living and nonliving worlds where the mode of reasoning was *inference* from general observable features and the apparent purpose reflected in those features.

118

The higher-level *design argument* today is based on observation of *actual* underlying electrochemical-mechanical *mechanisms* that are functional in biology and reveal how things work down in the fundamental machinery of the biological cell where the life processes themselves take place. Familiar examples of these features are the information content of DNA, the genetic code and its function, the process of transcription, protein synthesis, the manufacture of molecular machinery, and the construction of what can only be described as marvels of micro-engineering design. The Design Hypothesis in biology is not *imposed* as an *a priori* assumption: it *emerges a posteriori* as an inference from direct scientific observations. Logically and scientifically, this is as solid as it gets.

The information in DNA not only *looks* like information, it functions as information—it *is* information. The genetic code not only *looks* like a code, it functions as a code—it *is* a code. The molecular machinery within the cell not only *looks* like machinery, it functions as machinery—it *is* machinery. The notion of *apparent* design collapses under the weight of these fundamental scientific facts. Design is *present and operative* throughout the natural world.

> *Design is part of nature in precisely the same way that Law is part of nature. Both are present and operative in the natural world, and both are known by the observable effects they produce.*

❧ ❧ ❧

The modern version of intelligent design can be stated as follows:

(1) Intelligent design is the *hypothesis* that certain features observed in the natural world cannot be explained solely in terms of the laws of physics and chemistry, chance, and time, and that these features exhibit properties commonly attributed to design. Important design indicators are the appearance of some combination of specification, information, an abundance of low probability events and arrangements (complexity), purpose, and function in natural systems. The design hypothesis is established through logical inference from direct scientific observations.

(2) A basic tenet of the "design hypothesis" is that design is empirically detectable in nature through the methods of science. If science can detect "law" in nature, it can also detect "design" in nature.

(3) While intelligent design may have implications for a transcendent designer behind nature, it does not presuppose such a designer or depend on belief in such a designer.

While intelligent design is loosely referred to as a *theory,* its actual status scientifically is determined by its place in logical structure. Accordingly, ID is no more than a *hypothesis.* Abundant examples are found in the design inferred from direct scientific observation of features, operations, and functions within the biological cell that have the appearance of having been designed for a purpose. The design hypothesis is established by the design inference. Design in nature is identified through the same reasoning process used in identifying law in nature: inference from directly observed evidence.

While the biochemical composition of the cell can be explained by the normal laws of physics and chemistry, there are *higher-order* processes at work within the cell that cannot be explained in terms of those laws alone. Such higher-order processes include the storage, processing, and use of complex specified information, data transcription and editing, regulation, feedback, control, repair, highly coordinated and integrated communication and transportation systems, assembly of sophisticated molecular machines, energy processes to drive cellular operations, and a multitude of other processes and operations where it is the overall cellular structure, component designs, and system organization that integrates this maze of processes.

The design and function of the biological cell reflects all of the properties that are common to the fields of systems engineering and information technology where it is human engineers who design such systems. The primary difference between systems designed by human engineers and the design of the systems that are found in the biological cell is that the complexity and efficiency of the biological systems exceed the ability of human engineers by a degree that can only be described in astronomical terms—and at a microscopic level. These observations pose absolutely staggering challenges to both a naturalistic account of the origin of life and a strictly Darwinian account of the history and diversity of life.

The idea of a *code* has to do with an abstract nonphysical conception of the mind that can be implemented by intelligent agency in systems that store and use information based on that code for purposes determined beforehand by that agency and reflected in the overall design of the system

121

within which it resides and functions. But how does a code get impressed on a biochemical system without the aid of intelligent agency, and how does information based on that code get impressed on matter? How do chemicals and molecules come to "know" the code? No one knows. It is simply a huge mystery. Further, what accounts for the extraordinary efficiency and reliability of nature's code where computer scientists can design millions of different coding systems that might accomplish the same functions and serve the same purposes, but none even begin to approach the efficiency and reliability of nature's code?[30]

Then there is the nasty problem of the protein *chicken-and-egg* dilemma where molecular machines within the cell are required to assemble proteins, but the machines themselves are constructed from proteins. How such systems might arise through unguided natural processes alone is a mystery. Such difficulties run throughout biology.

Intelligent design as it pertains to the biological sciences is not a comprehensive theory of biological origins as is the case for neo-Darwinism in that it does not provide a historical narrative regarding the history of life and the origins of mankind. ID makes no speculations about unobserved events and processes of the remote past and is not dependent on metaphysical presuppositions about the existence or nonexistence of a transcendent designer. Intelligent design in biology is simply an inference that arises logically and naturally from direct scientific observations regarding various aspects of cellular and molecular biology.

ID makes very modest claims, never loses sight of the evidence, and does not proceed beyond the design hypothesis. *Darwin* and *design* look at the same evidence, ask the same questions, and relate to that evidence in two radically different ways.

> ***Design is inferred from the evidence. Darwinism is
> presupposed and imposed on the evidence. Design is
> established inductively. Darwinism is established
> deductively. Design is science. Darwin is dogma.***

ID simply sets forth the *hypothesis* that the complex biochemical systems observed in biology associated with the life processes reflect the work of intelligent agency, suggesting that design methods and information technology can be applied in understanding the operation and behavior of biological systems. And, as described before, these methods have been brought together to form the field of *systems biology*. It is more than a little ironic that committed Darwinists must use design principles to explain biological systems they sincerely believe are the result of chance processes.

As distasteful as this might be to Darwinists, students must be exposed to both the evidence for and the evidence against biological evolution as well as both the evidence for and the evidence against intelligent design. Given a proper exposure to the evidence, the astute student will likely come to the conclusion that neo-Darwinism has both evidentiary and methodological flaws. The learning environment provided in the public school science classroom must accept such a possibility both in principle and in practice. The problem is that it is those evidentiary and methodological flaws that public education deliberately conceals from students (and everyone else) to protect the Darwinian dogma.

In good Socratic form, questions concerning fundamental aspects of biology should be asked, such as: "What is the source of the genetic code, and how did it get impressed on matter? What accounts for its amazing efficiency? How did nonliving chemicals come to "know" the code? Was it by chance or by design? If the assembly of proteins requires molecular machines made up of proteins, how did this most fundamental of all chicken-and-egg dilemmas get resolved? Was it by chance or by design? What was the source of the information in DNA, and how did the extraordinary molecular machinery and processes that regulate and integrate cellular operations come to be? Was it by chance or by design?"

These are the truly interesting and challenging questions that are currently unresolved in biology. However, within the ruling paradigm of neo-Darwinism, publicly you cannot ask such questions because of the rule: "You do not question Darwin."

    ❦   ❦   ❦

Deep within the evolutionary establishment, the inadequacy of the current evolutionary paradigm in explaining the existence of biological complexity and macroevolutionary processes is openly admitted. The crisis in evolutionary biology publicly announced by Michael Denton in 1986, after over thirty years, has not been resolved. It has only been through the actions of the US Supreme Court in *Edwards* in 1987 that the position of biological evolution in public education has been secured. Evidence of the gravity of this crisis is revealed through recent technical meetings and publications, all of which in one way or another question the adequacy of current evolutionary theory. These concerns however, are expressed in relation to the *mechanics* of evolution, not the *fact* of evolution, which is presupposed.

In 2011, University of Chicago professor James Shapiro published his book *Evolution: a view from the 21st century*.[31] In 2012, New York University Professor Thomas Nagel (previously discussed) published his book *Mind & Cosmos*.[32] In November 2016, the Royal Society met in London[33] to discuss problems with current evolutionary theory, calling for revision of the standard theory. In 2017, two fine books were published by highly regarded mainstream scientists, *Dance to the Tune of Life*[34] by Denis Nobel, emeritus professor at the University of Oxford (who attended the Royal Society meeting in 2016) and *Purpose & Desire*,[35] a wonderful book by J. Scott Turner, professor of biology and physiology at State University of New York. While a naturalistic explanation of evolution is still sought, there is a growing conviction that Darwinism is not the answer, and there is no ready replacement.

In the meantime, in public education, it is business as usual.

✑   ✑   ✑

Public education must accept that the time is rapidly approaching when the people of the communities it serves will no longer tolerate the narrow dogmatism that has been institutionalized in the teaching of evolutionary biology for the last thirty years. I suggest that content from this present chapter should provide some basic guidelines and key ideas to consider in formulating educational policy and curricula materials that are set within a more objective philosophical, scientific and educational framework. Abundant resources in support of that end are available at the Discovery Institute[36] and Access Research Network[37].

CHAPTER 8

THE DARK SIDE OF DARWINISM

The impact of Darwinism on Western culture in the twentieth century was profound, reaching far beyond the fields of science and biology. It exerted dramatic influences on religion, forms of government, political and legal theory, and the course of history. Much wind is expended, exalting the wonders of Darwinism, but wind is about all we get when we catalog its real benefits to mankind. What we do not hear about is the horrors that were unleashed on mankind in the twentieth century in the name of Darwinism.

In 1859, Charles Darwin's book *On the Origin of Species* introduced the world to his theory of natural selection, the force, he claimed, that drives biological evolution. Through the process of the sampling of life's random variations—as the story goes—nature, over and over and over again, picked the "fittest" of those variations, and in so doing slowly transformed the "simple" single-celled organisms that existed a billion years ago into the complex and wondrous creatures of today's world. Of course at that time, no one—including Darwin—had any idea as to the actual complexity of even the simplest single-cell organism. The truth is that Darwin's theory was largely based on speculation and a profound ignorance of the true nature of biology and the complexity of life. There was no real convincing evidence in its support, and it was of little or no practical use. Its real value lay in its ability to provide the basis for a new secular religion to replace Christianity, which, as a result of scientific progress, was seen by many as becoming increasingly irrelevant to life in late nineteenth-century England. Today, in spite of its inability to provide anything more than the framework for "just-so" Darwinian fairy tales, Darwinism reigns supreme *in* science, not *as* science, but as naturalistic ideology.

Darwin's intent in publishing *Origins* was that his theory be accepted, not just by the scientific world, but by the masses. However, he had a problem. It was one thing to say that the fishes of the sea, the birds of the air, and the beasts of the field evolved, but it was another thing altogether to say that the same was true for mankind. This would surely offend the Christian ladies and gentleman of Victorian England who believed that their existence was the result of the special creative work of God and saw themselves as children of God, not descendants of apes. Darwin thought it best to let everyone get a little more comfortable with these ideas before exposing them to the disconcerting "truth" about the origin of mankind, much less its dark and troubling implications. All of this would have to wait until later.

Thus, in writing *Origins,* Darwin reluctantly stopped short of claiming that mankind, along with the rest of the living world, was also the product of natural selection, though that was his ultimate intent. In the last sentence of his first edition (omitted in later editions) he even gratuitously threw in something for the Christians:

There is grandeur in this view of life, with its several powers, having been originally breathed into a few forms or into one; and that, whilst this planet has gone cycling on according to the fixed law of gravity, from so simple a beginning endless forms most beautiful and most wonderful have been, and are being, evolved.

Charles Darwin had a cousin named Francis Galton. Galton had an excellent understanding of Darwin's work, but he had little concern about British religious sensibilities. Neither was he married to a wealthy and devout Christian lady as was his cousin Charles. In 1869, Galton published a book, *Hereditary Genius*,[1] based on Darwin's *Origins* where he not only made it clear that mankind descended from lower forms, but he also foresaw the emergence of a new "science" based on Darwin's theory.

Galton was well aware that one of Darwin's strongest arguments supporting natural selection came from domestic breeding. He reasoned that these same principles could be applied to human populations for the purpose of eliminating undesirable traits, just like breeders did with domestic animals. This reasoning provided the basis for a new social science which Galton called "eugenics."

In *Hereditary Genius*, Galton cleared the way for Darwin with respect to any remaining religious issues. Darwin then published his second work, *The Descent of* Man,[2] and it was in this book that he explicitly laid out the foundations of eugenics. Darwin writes:

With savages, the weak in body or mind are soon eliminated; and those that survive commonly exhibit a vigorous

128

state of health . . . We civilized men, on the other hand, do our utmost to check the process of elimination; we build asylums for the imbecile, the maimed, and the sick; . . . and our medical men exert their utmost skill to save the life of every one to the last moment . . . Thus the weak members of civilized societies propagate their kind. No one . . . will doubt that this must be highly injurious to the race of man.

If . . . various checks . . . do not prevent the reckless, the vicious and otherwise inferior members of society from increasing at a quicker rate than the better class of men, the nation will retrograde, as has occurred too often in the history of the world.

In *Origins,* Darwin introduced his theory of natural selection. In *Descent,* he presented his arguments for the evolutionary origins of mankind and eugenics. At the end of the nineteenth century, progressive utopian thinking was in the air among the elite in Europe and America. This combination of ideas set the stage for the twentieth century and some of mankind's darkest times.

๑๕ ๑๕ ๑๕

For three decades following its introduction in 1869 by Francis Galton, eugenics existed as little more than a somewhat bizarre topic of discussion in social and academic circles.[3] It likely would never have progressed much further than that had it not been for extensive financial support provided by philanthropies such as the Carnegie Institution and the Rockefeller Foundation that funded the eugenics research of some of America's most respected scientists at such prestigious universities as Stanford, Yale, Harvard, and Princeton. During the early decades of the twentieth century, eugenics gained the support of our nation's social, political, and academic elite, being championed by progressives, such as Woodrow Wilson, Margaret Sanger, and Oliver Wendell Holmes.

Even the United States Supreme Court endorsed eugenics. In the High Court's infamous 1927 decision in *Buck v. Bell*, Justice Oliver Wendell Holmes wrote, "It is better for all the world, if instead of waiting to execute degenerate offspring for crime, or to let them starve for their imbecility, society can prevent those who are manifestly unfit from continuing their kind . . . Three generations of imbeciles are enough."

Years later, the Nazis at the Nuremberg trials would quote these very words of Holmes in defense of their social "cleansing" policies.

Following the Supreme Court's decision on eugenics, cruel laws were passed in twenty-seven states, opening the floodgates for thousands to be coercively sterilized or otherwise persecuted as "subhuman." The victims of this movement were poor whites, blacks, Jews, Native Americans and Mexicans, epileptics, alcoholics, petty criminals, homosexuals, the mentally ill, and just about anyone else who did not conform to the concept of a white, blond, blue-eyed Nordic "master race." Master race? Yes, that's right. The concept of the Nordic "master race" did not originate with Hitler's Nazi Germany. It came from the United States and was cultivated in California's universities decades before Hitler came to power.

US and German eugenic scientists enthusiastically collaborated in their work, and while it was clear that the Americans were leading the way in research, soon it was the Germans who led the way in large-scale applications of that research as social policy where eugenics was written into German national socialism, known as the Nazi Party. American scientists vigorously approved the progress in Germany and lamented their own inability to do the same in America. The Third Reich persecuted the Jews, and American eugenic scientists cheered. They forcibly sterilized over 200,000 individuals (after which they quit counting) deemed to be "unfit," and American eugenic scientists cheered. They gassed 100,000 aged, mentally disabled, and otherwise "useless" people, and still American eugenic scientists cheered. It was not until the massive genocide of the Jews and the hideous eugenic medical experiments performed by Dr. Josef Mengele at Auschwitz (who incidentally, was funded by Rockefeller before he went to Auschwitz) were revealed at Nuremberg that the American eugenic scientists were finally silenced.

❧ ❧ ❧

The connection between Darwin and Hitler is hotly debated. Darwinists would like to distance themselves from Hitler and certainly do not want Darwin implicated in Hitler's dirty work. But the logical pathway from Darwin to eugenics in its implementation in both the United States and Germany (and England as well) was direct and simple—just a matter of connecting the dots. In Germany, having embraced the American vision of the Nordic master race, the transition from eugenics to genocide was just a matter of making the connection to one more dot—and then the killing began.

Richard Weikart, professor of history at California State University, Stanislaus, in his book *From Darwin to Hitler*,[4] explains how deeply entrenched Darwinian thinking was in Germany by the time of the First World War and how it progressed in the Second World War to its logical conclusion in the Holocaust. This is intellectual history that Darwinists would like to forget. It is also intellectual history we absolutely cannot afford to forget.

If public schools are going to teach Darwinism, they should also inform students of its rotten fruits. As Weikart points out, without Darwin, Hitler would not have had the "scientific" justification to exterminate six million Jews. Darwin gave him that so-called "scientific" justification.

∽ ∽ ∽

It is difficult for us today to understand how such a contemptible practice as eugenics could be given legal sanction by the US Supreme Court and enacted into laws in over half the states in our nation. We ask, "How could this happen in America, a Christian nation?" It is one thing to take reasonable measures to improve domestic stock, but what social or moral principle could justify a similar program for humankind where fundamental human rights to life and liberty were trumped by government mandate? The answer lies in understanding the mind of the early twenti-eth-century progressive thinkers.

From the perspective of the elite progressives of that era, it was a time like no other time in human history. It was the dawning of a new age. Mankind, in his long journey down through the ages, had finally come to an understanding of his true nature, and the fountain of that understanding was science. For the first time in history, Man knew who he really was. The work of the Enlightenment, inspired by reason and scientific discovery, had freed mankind from the shackles of ignorance and religious superstition. It was the enlightened thinkers alone who fully understood the significance of this *Great Liberation* wrought by the hand of science, and with this new vision of the future, they would lead mankind into the utopian promised land.

Until that moment in history utopia had only been a distant dream. Now it was on the horizon, it was within sight, and the path was clear. Darwin had made the greatest discovery in history, and from that discovery emerged the principle of eugenics, and through eugenics, mankind would attain his destiny of evolutionary perfection, a biological utopia, not in millions and millions of years, but within a few generations. They had seen the future, and that future was white, blond, and blue-eyed—the Nordic master race. It was time, using the science of eugenics, to put the finishing touches on humanity.

They had no idea that they were unleashing an outpouring of evil on mankind such as had never before been seen in all of human history. This was a dark and shameful moment in the history of mankind. But it did provide an unforgettable answer to the single most important question the eugenicists never asked. What happens when you uproot science from a guiding moral philosophy? You get Auschwitz.

◈ ◈ ◈

We have known since Nuremberg what happened when eugenics came to Germany. But only recently through the work of outstanding journalist Edwin Black[5] in his book *War Against the Weak* have we come to understand what happened when eugenics came to America. We may wonder just how far the eugenics program in America might have gone if it had not been derailed by collateral damage that occurred when the Allies took down the Third Reich in Germany. What was their agenda, and how did they intend to advance that agenda? The vision of a white, blond, and blue-eyed future was well established in the American universities, but how do you implant that vision in the culture? Easy—introduce eugenics ideology (under the pretense of science, of course) into the curriculum of public schools, and through that means, the eugenics message would spread throughout American culture. But could such a thing actually happen in America? Consider these excerpts from an early-twentieth-century biology textbook[6]:

> Improvements of Man.—If the stock of domesticated animals can be improved, it is not unfair to ask if the health and vigor of the future generations of men and women on the earth might not be improved by applying to them the laws of selection.

> Eugenics.—When people marry there are certain things that the individual as well as the race should demand. The most important of these is freedom from germ diseases which might be handed down to the offspring. Tuberculosis, that dread white plague which is still responsible for almost one seventh of all deaths, epilepsy, and feeble-mindedness are handicaps which it

is not only unfair but criminal to hand down to posterity. The science of being well born is called eugenics.

Parasitism and Its Cost to Society.—Hundreds of families such as those described above exist today spreading disease, immorality, and crime to all parts of this country. The cost to society of such families is very severe. Just as certain animals or plants become parasitic on other plants or animals, these families have become parasitic on society. They not only do harm to others by corruption, stealing, or spreading disease, but they are actually protected and cared for by the state out of public money. Largely for them the poorhouse and the asylum exist. They take from society, but they give nothing in return. They are true parasites.

The Remedy.—If such people were lower animals, we would probably kill them off to prevent them from spreading. Humanity will not allow this, but we do have the remedy of separating the sexes in asylums or other places and in various ways preventing intermarriage and the possibilities of perpetuating such a low and degenerate race. Remedies of this sort have been tried successfully in Europe and are now meeting with success in this country.

The name of this biology textbook was *A Civic Biology* by George William Hunter, the textbook that was in use in Dayton, Tennessee, in 1925 in the famous Scopes trial. Public schools at that time were, under the pretense of science, indoctrinating American school children into the same eugenic ideology that was being exported to Germany.

∞   ∞   ∞

Today you will not find anything about eugenics in public school biology textbooks, but its foundational principles, in the form of evolutionary biology, are still solidly entrenched in the biology classroom. Darwin, through his naturalistic account of the emergence of man, provided the scientific basis for the naturalistic worldview and within this framework of thinking, eugenics as social policy is easily justified—as long as you do not have a moral philosophy to interfere.

It should be deeply disturbing to Christians that over the years, many with sincerely held Christian beliefs endorsed slavery, and as soon as we finally got slavery out of our blood, we turned to eugenics, and soon as we finally got eugenics out of our blood, we turned to abortion (which is just another form of eugenics). Maybe someday we will get abortion out of our blood. Who's next? The elderly?

Eugenic practices in America were finally ended, not following the Nuremberg trials in 1945–46, but in the 1970s—but the story does not end here. The Hitler fiasco may have given eugenics a bad name, but the eugenic ideology of racial perfection lives on in America. It just has a new name: human genetics. All of the original eugenics organizations and journals are still around. They just changed their names.

The modern science of genetics and the mapping of the human genome open up all kinds of interesting opportunities for application of eugenic principles to human society. For instance, Nobel laureate Francis Crick said, as quoted by *Pacific News Service* in January 1978:

> No newborn infant should be declared human until it has passed certain tests regarding its genetic endowment. If it fails these tests, it forfeits the right to live.

Who would render the test, grade its results, and carry out its sentence? The *state,* of course. In a progressive socialistic utopian system, the existence of a "defective" individual—referred to as "human waste" back in the good old days of eugenics—will become a burden on the state and the society that supports the state, and of course, that would be "unfair" to all the "perfected" members of a fully realized utopian society.

The final chapter of eugenics in America has not been written. We should be careful to suppose that a society that has learned to live with abortion (approaching 60,000,000) will not also learn to live with other less offensive and far more practical and economically beneficial practices in dealing with or disposing of "human waste" in our society, especially if it is "only fair" and for the "greater good." We can expect that eugenics in some form (or disguise) will eventually play an important role in a fully realized secular state undergirded by progressive political ideology.

✍ ✍ ✍

The eugenics movement had its birth in England, and it is quite natural that it would have a strong influence there. Nowhere is Darwin more honored than there. Few achieve the distinction of being buried in

Westminster Abbey where Darwin was laid to rest next to the eminent scientist Sir John Herschel and not far from Isaac Newton. On the occasion of the 150th anniversary of Darwin's publication of *On the Origin of Species* in 2009, British broadcaster and author Dennis Sewell published his book, *The Political Gene*,[7] telling us the story of eugenics. It is a particularly good read, but far more than that, it shows how eugenics was institutionalizing barbarism in the name of social Darwinism in modern political philosophy. Sewell tells us how eugenics emerged in England as a "Darwin family business" and how it quickly found a place in English society through the dire warnings of scientists about the deterioration of the gene pool. Eugenics was called in to take control of the "problem," somewhat like climate scientists today are issuing dire warnings, trying to save us all from global climate change. The principle impact of Darwinism on the world was not scientific, it was political. Nevertheless, due to a great public relations program, Darwin is a hero of Great Britain.

In the last chapter of his book, Sewell provides extraordinary insight into the profound and destructive aftermath of Darwinian social policies in the twentieth century[8]:

> We have been living off the accumulated cultural capital of our Christian past to sustain not just our ethical and moral precepts but our political values as well.

> The Darwinist ideology simply cannot supply the moral and ethical basis for the democracy we want to live in.

It took nearly a century to sort out the chaos left by Marx and Freud, and to bring their ideas back within reasonable intellectual bounds; but the task of clearing up after the great revolutionary thinkers of the nineteenth century will not be complete until Darwin has been put back in his box.

Well said, Mr. Sewell. Well said. We're working on it.

CHAPTER 9

DARWIN, DESIGN, AND THE PHILOSOPHERS

F rom antiquity, most cultures have embraced some notion of a designing intelligence behind nature. The ancient Greek philosophers saw the rational order and regularities of the natural world. They marveled at the purpose reflected in the wonderful harmony between living creatures and the nonliving world and reasoned that all of these things could not have come into being apart from a designing intelligence. This reasoning gave rise to the *teleological* view of nature, the idea that nature was designed to serve some intended end, and that view has been part of Western thinking ever since. The Greek stoics, though not informed by revelation, took their cues from the natural world and, through reason, presumed a God of nature who was ruler over the universe and a universal moral law (natural law) impressed upon the heart of man. While these characteristics provided the Greeks with the basis for the teleological view, they were insufficient in inferring a transcendent being anything like the knowable God of the Bible.

Natural law has also played an important role in Western civilization, being affirmed by thinkers such as Cicero, Thomas Aquinas, and John Locke and coming to full fruition in America through our Founding Fathers when Thomas Jefferson wrote its basic principles into the Declaration of Independence. Our right to exist as a free nation was established under the "Laws of Nature and of Nature's God," and as citizens of this new nation our individual rights to life and liberty, our *natural rights*, are given by our Creator, not the state. In the eighteenth century, the teleological view of the world and natural law were not only relevant to natural philosophy, they were fundamental to the American founding.

Jefferson had an opinion on the idea of design in nature as well, where in his later years he expressed his views in a letter[1] to John Adams:

> I hold (without appeal to revelation) that when we take a view of the Universe, in its parts general or particular, it is impossible for the human mind not to perceive and feel a conviction of design, consummate skill, and indefinite power in every atom of its composition. . . . So irresistible are these evidences of an intelligent and powerful Agent that, of the infinite numbers of men who have existed thro' all the time, they have believed, in the proportion of a million at least to Unit, in the hypothesis of an eternal pre-existence of a creator, rather than in that of a self-existent Universe.

Thomas Paine, merciless critic of organized religion that he was, in his lecture "On the Study of God" delivered in Paris in January 1797 to the Society of Theophilanthropists,[2] summarized his thoughts on the relationships between God, science, and public education. According to Paine, God is *Author* of the principles of science and should be properly acknowledged as such in the schools. He said:

> It has been the error of the schools to teach astronomy, and all the other sciences and subjects of natural philosophy, as accomplishments only . . . whereas they should be taught with reference to the Being who is the author of them: for all the principles of science are of Divine origin. Man cannot make, or invent, or contrive principles. He can only discover them; and he ought to look through the discovery to the Author . . . The evil that has resulted from the error of the schools in teaching natural philosophy as an accomplishment only has been that of generating in the pupils a species of atheism.

That was true in late eighteenth-century France, and it is true as well in twenty-first-century America. In France, even the deist Voltaire (1694–1778), one of the more extreme anti-Catholics of the Enlightenment philosophers, feared that the mechanical view of the universe, if taken to the extreme, would push God out of nature, thus eliminating any rational basis for morality. For Voltaire, design in nature was a given. He said:

> All nature cries aloud that there is a supreme intelligence.[3]

Voltaire also believed that the existence of God was as much a matter of reason as of faith.

What is faith? Is it to believe that which is evident? No. It is perfectly evident to my mind that there exists a necessary, eternal, supreme, and intelligent being. This is no matter of faith, but of reason.[4]

Teleology and natural law are two of the great legacies of Western civilization. Both are inferred from the natural world alone, yet they transcend the natural world. Both were conceived during the classical Greek period, and both were borne down through the ages by Christianity. However, in the twentieth century, a movement emerged whose objective was to systematically eliminate both from Western culture as relics of the prescientific age, unsuitable for an enlightened people who through science and reason have been liberated from the superstitions and bonds of religion. Darwinism lies at the heart of the principle scientific argument against both.

ຈ ຈ ຈ

The writings of three eighteenth-century philosophers—the Scot, David Hume (1711–1776), the German, Immanuel Kant (1724–1804), and the Englishman, William Paley (1743–1805) had much to do with shaping the debate concerning the teleological view and the place of design in nature. Newton had published his *Principia* in 1687, and by the time these three philosophers "came of age" intellectually around mid-eighteenth century (give or take a decade or two), the *mechanical philosophy* was broadly accepted as a sort of "benchmark" for guiding scientific inquiries into the nature of the world. Newton was enormously successful in developing an understanding of mechanics based on law. Why would we not expect that there might be such a law that explains life?

Rene Descartes in the century before had posited that reality had a dual nature, being comprised of two fundamental components: mind and matter. In examining the world of living and nonliving things, we find that in special cases (such as animal life), when mind occupies matter, you get life, and in this union of mind and matter, matter serves the needs of mind. Mind runs the show. Descartes, as a Christian, presupposed that Mind came first, then matter.

Newton showed that the behavior and operation of the nonliving world, that is, the *mechanics* of the nonliving world, is explained solely through natural *law*. But "law" is an immaterial conception of *mind*, and its function in the material world is to regulate the behavior and operation of that world in a prescribed manner. Where did these laws and their prescription come from, and how did they take on their particular form? Such properties of the material world makes perfect sense from the teleological perspective where all of the evidence suggests that it was the creative work of a transcendent guiding intelligence, a *Law Giver*, that accounts for what we observe. The naturalist, however, because of his view of the world and his abhorrence of design must somehow account for law apart from mind, and the only non-metaphysical alternative to mind is *chance* and lots and lots of time. From the perspective of the naturalists, mind emerged from matter. Matter came first, then mind. It was all an enormous cosmic accident. There is no need to pursue the question any further; otherwise, you might get "confused," you know, like the Christians who are the most "confused" of all.

Mind is simply a consequence of the properties of matter, even though no one has even begun to explain how that works. We will hear from Nobel Laureate George Wald on this subject a little later who, having staked his claim with Darwin, later changed his mind. How did that happen? He actually thought about it!

The place of mind in the natural order is fundamental, and any naturalistic explanation of the natural order, including life and mankind, that cannot account for the emergence of mind is hollow, incomplete, and without a foundation. It is when we begin to think deeply about such things that we see the rickety foundations of naturalism.

In considering how the naturalist might provide an explanatory account of the *living world,* we see absurdity taken to a new level. There is something fundamentally different about living things and within the world of living things there is something fundamentally different about man. Living things are made of the same nonliving "stuff" as the nonliving world, but they have fundamentally different properties. They are *alive,* which of necessity entails not only a material means for physically interacting with the world of material things but a material sensory system that connects the mind to the nonliving world (and to itself, for that matter) and makes that interaction possible. Given our makeup of nonliving "stuff" which is composed of nonliving *dead* molecules, how do you breathe *life* into those *dead* molecules? The naturalistic explanation for the "breath of life" is that somehow, when some minimum threshold of complexity is crossed, life emerges.

For the Christian natural philosopher, in spite of these limitations on their understanding concerning the living world, the "benchmark" set by Newton concerning the *material world* was also imposed (at least tentatively) on the *living world* as well under the presumption that there *must* be (or might be) some natural law or set of laws that God put in place that could account for the living world just as Newton's laws provided an account for the mechanics of the nonliving world. This "law-based" view of the living world arose in the eighteenth century and was prevalent well into the nineteenth century providing the scientific and philosophical backdrop for Darwin as he thought about these things. Accordingly, we see that Darwin's naturalistic hypothesis was well within the mainstream of evolutionary thought, even for Christian natural philosophers where he supposed, along with many others, that unguided material processes alone *might* be able to account for life.

Central to these debates was the question of *design in nature* and this was the question taken on by Hume, Kant and Paley.

ᕯ ᕯ ᕯ

Now, before we step into the world of Hume, Kant, and Paley, a disclaimer is needed. I am not a professional philosopher. I have not dedicated my life to the study of philosophers and their philosophies. Nevertheless, given modern scholarship which seems to be sincerely dedicated to an unbiased examination of what these three giants actually said and intended rather than to providing interpretations that support a dogma, it is not at all difficult to establish that their views have not always been faithfully presented to the popular culture, and that was particularly true for Hume. I believe that which I have written here is trustworthy for all, including Hume.

✅ ✅ ✅

Newton's mechanical philosophy provided the foundation for the philosophical works of Hume and Kant. Both Hume and Kant appealed to reason alone, both addressed the question of design in nature, and both, given the accepted standards for philosophical inquiry (mainly that of avoiding unnecessary assumptions), they both had an appreciation for the enormous conceptual gulf that separated the material world and our mental perceptions of that world. They were committed to positing their arguments within the mechanical view of the world, but it was impossible within that view for them to envision a *mechanism* that could account for a true correspondence between reality and the mind. Consequently, both would say that the order that we perceive in nature is imposed by our minds on otherwise meaningless sensory inputs, giving the appearance of an orderly world. Both Hume and Kant would say that we cannot really know for sure what is going on "out there."

This is hardly a satisfying explanation regardless of your preferred metaphysics. It is understandable why Hume would take such a view, given his religious skepticism, but it is puzzling why Kant would not (or could not) acknowledge that such correspondence is universally realized because that correspondence had been built into the universe by God from the first. Our mental "operating system" could not possibly process the enormous sensory input we receive every moment unless both it and our sensory systems were designed to provide that correspondence. Further, the jump from "neural impulse" (a physical phenomenon) to "thought" (a mental process) and "mental image" (a mental representation) cannot possibly be a strictly mechanical process because the end product is not mechanical. Rather it is a mental representation of some aspect of the material world. Regardless of our metaphysics, we all have to admit that therein lies a great mystery.

Hume's upbringing and education were centered on preparing him for a life guided by a strict Scottish Calvinist religious tradition. He studied at both Edinburgh University and in France and read broadly in history, philosophy, mathematics, and the natural sciences. By the time he was in his early twenties, he had rejected the Christianity of his childhood, developed a life-long antagonism toward organized religion, and eventually chose the life of a "scholar and a philosopher." While his works have been used to serve the cause of naturalism, it is not clear that he was in fact an atheist. You must judge for yourselves, based on what he had to say, and momentarily, I will tell you what he had to say. Actually, it was quite remarkable.

Hume[5,6,7] wrote as a skeptic, carefully framing what has been widely recognized as a devastating critique of *design* and is often credited with having nailed the coffin shut on the design argument. His arguments followed the traditional disputational method of the scholastics of the Middle Ages, designed to uncover and establish truth in theology and the natural sciences. Following this method, Hume made the strongest case possible for the existence of a designer behind nature and then proceeded to tear it down. His attacks against the rational basis for Christian faith were seen as devastating by scholars of the eighteenth and nineteenth centuries. The positivist epistemology and naturalistic philosophy that came to dominate Western culture in the twentieth century had their roots in part in Hume's thinking. Darwin was strongly influenced by both Hume and Kant, and all three in their own way contributed to making the twentieth century the bloodiest century in history.

Kant[8,9] was raised within a reform movement of the Lutheran church that favored inner religious conversion and upright conduct over formalized doctrine and liturgy. He remained absolutely faithful throughout his life to that upbringing in all of his works. Kant said that it was the works of Hume that "awakened him from his dogmatic slumber," and once aroused, he devoted much of his life to refuting Hume's philosophical program, a program that challenged not only the rationality of Christian faith, but the rationality of Newtonian mechanics. Hume challenged and Kant defended a central truth. If there was no fundamental basis for Newton's laws in the material world, then neither could there be a fundamental basis for God's laws concerning the affairs of men.

Much of Kant's works was viewed as being critical of religion, and to some extent this was true; however, the negative elements of Kant's philosophy of religion should not be taken as denials of, or even challenges to, faith. Rather, Kant set forth these arguments in order to ensure that we do not allow *reason* to be used abusively such that it undermines the legitimate place of *religion* in our lives. Hence, despite more than two centuries of commentators who have held that Kant's criticisms were expressions of hostility toward religion, his critique was not meant to abolish faith but to preserve it. Modern scholarship has exposed this historical abuse of his works. For Kant, man was imbued with the ability to perceive both spiritual and material realities where the "moral instinct" was just as much a part of reality as the laws of mechanics. Kant, exercising extraordinary insight, took the universal "moral instinct" shared by all of mankind as an *objective* rather than *subjective* aspect of reality. Individuals may have their own subjective response to this instinct, but the instinct was the same for all. For Kant, this instinct was as much an indicator of God's works in man as His laws of mechanics were an indicator of His works in the natural world. His point was compelling.

Kant was sympathetic to the design argument, describing it as the "the oldest, the clearest, and the most accordant with the common reason of mankind." He further regarded it as having considerable utility for the natural sciences. Kant claimed that the design argument *succeeds* in at least establishing "an *architect* of the world" and a cause "proportioned" to the order of nature. He wrote, "We have nothing to bring against the rationality and utility of this procedure, but have rather to commend it further." What Kant could not accept, however, is the leap from a "wise author of nature" to an "infinite Creator." When the argument moves from *architect* to *creator,* it proposes an "original" and "supreme" cause, and in so doing, it calls for a being whose existence depends upon nothing but itself. The designer, for instance, might be a dependent agent, like Plato's demiurge. Kant rightly noted that the capacity to create something from nothing entails abilities that greatly exceed the capacity to establish order and impose design. He says nothing, however, against the idea that the creator of the world and the designer of the world might be one and the same, which to many (myself included) seems self-evident. No doubt, it was also apparent to Kant, however, within the rigorous limits of reason prescribed by his assumptions as a philosopher, he was not allowed to make that connection.

Kant's commitment to the mechanical view was reflected in his hope that "one day there would arise a second Newton who would make intelligible the production of a single blade of grass in accordance with the laws of nature the mutual relations of which were not arranged by some intention." It is apparent that Kant held soundly to the "creation by law" or "greater God" hypothesis; however, it is curious that he would hope that such a law would not reflect intention. Nevertheless, his dream of a mechanical explanation of life has thus far been unrealized, and given what we know about biology today, it will never be realized.

In the wake (or aftermath) of Hume and Kant, William Paley, Anglican clergyman and fellow of Christ's College, Cambridge, forcefully and elegantly elucidated the notion of a transcendent designer behind nature in his influential 1802 exposition *Natural Theology*[10] in which he presented his well-known argument for design based on one of the most famous metaphors in the philosophy of science, the *watch* and the *watchmaker*. After the huge intellectual efforts bought to bear on the question by Hume and Kant, Paley set forth a classic design argument that stands on its own, in a sense putting in the last word in the debate over teleology and design. Here is Paley:

> In crossing a heath, suppose I pitched my foot against a *stone,* and were asked how the stone came to be there; I might possibly answer, that, for any thing I knew to the contrary, it had lain there for ever: nor would it perhaps be very easy to show the absurdity of this answer. But suppose I had found a *watch* upon the ground, and it should be inquired how the watch happened to be in that place; I should hardly think of the answer which I had before given, that, for any thing I knew, the watch might have always been there. Yet why should not this answer serve for the watch as well as for the stone? why is it not as

admissible in the second case, as in the first? For this reason, and for no other, viz. that, when we come to inspect the watch, we perceive (what we could not discover in the stone) that its several parts are framed and put together for a purpose, *e. g.* that they are so formed and adjusted as to produce motion, and that motion so regulated as to point out the hour of the day; that, if the different parts had been differently shaped from what they are, of a different size from what they are, or placed after any other manner, or in any other order, than that in which they are placed, either no motion at all would have been carried on in the machine, or none which would have answered the use that is now served by it.

Paley continues with a discussion of the springs, chains, wheels, the balance, pointer, and other parts of the watch, as well as their interaction and function. He then ties it all together:

The inference we think is inevitable; that the watch must have had a maker—that there must have existed, at some time and at some place or other, an artificer or artificers who formed it for the purpose which we find it actually to answer, who comprehended its construction and designed its use.

Materialists, however, while they cannot deny evidence of design in nature, claim that such evidence is insufficient for proof of the existence and action of a transcendent creator behind nature. They have a point, but that point can never rise above the level of speculation until they can show *scientifically* rather than through "just-so stories" that the origin of life can be attributed to natural laws and mechanisms acting alone as, in the terminology of the mechanical view of nature, *part of the overall "machine" acting without intention.*

Given the popularity of *natural theology* during this time, Hume's works were poorly received, and Kant's response was misrepresented. Both Hume's and Kant's works were, however, read by Charles Darwin, and they strongly influenced his ideas about evolution. Hume became more widely read in late nineteenth century when agnostic Thomas Huxley (1825–1895) began his crusade, promoting both Hume's and Darwin's works. Kant's works, however, were read and interpreted through the lens of Enlightenment naturalists and used to discredit the teleological view of nature. Again, modern-day scholarship exposes the biases that for two centuries misrepresented Kant's views.

In Hume's last work, *Dialogues Concerning Natural Religion* (published posthumously three years after his death) in which he launched his most devastating attack against religious faith, Hume leaves us with what might be considered to be his "parting shot" to the world concerning design in nature. In the end, Hume had to admit that it remained that there is something about the world that cannot be explained by natural causes alone. While it is apparent that his intent in the *Dialogues* was destruction of the teleological view of nature, it was clear to Hume that even after his devastating attacks, its place in nature could not be denied. His strawman was made of something far more durable than straw.

This final message from Hume comes to us through the fictional narrator of the *Dialogues,* Pamphilus, whose job it was to report the discourses of the three fictional participants, Cleanthes (who supports the teleological argument), Philo (the skeptic and critic of the design argument), and Demea (who attacks Cleanthes' teleological views). All three believe in God but advocate different understandings of his nature and, in particular, his relationship to the world.

In his final discourse, Philo (who most scholars believe speaks for Hume) said that "If . . . the cause or causes of order in the universe . . . bear some remote analogy to human intelligence . . . [then] what can the most curious, thoughtful, and religious man do except give a plain, philo-sophical assent to the proposition, as often as it comes up, and believe that the arguments on which it is based outweigh the objections against it?" A few sentences later, Philo says "To be a philosophical skeptic is to a man of letters, the first and most essential step towards being a sound believing Christian." This was followed shortly by Pamphilus' closing words:

> Cleanthes and Philo did not pursue this conversation much further; and as nothing ever made greater impression on me than all the reasonings of that day, so I confess that on carefully looking over the whole conversation I cannot help thinking that Philo's principles are more probable than Demea's, but that those of Cleanthes approach still nearer to the truth.

Scientific discovery during the last 100 years in the cosmological and biological sciences confirms Pamphilus' conclusions in ways he could not have imagined.

It is worth noting the positive things that Hume had to say about design in nature. In fact what he had to say was quite extraordinary and inspiring. It seems that if his intent was to construct a strawman and then tear it down, he tried way too hard, not in tearing it down, but in its construction. That strawman was quite durable. It still stands today.

In the introduction to his earlier work, *The Natural History of Religion*, Hume wrote:

> The whole frame of nature bespeaks an intelligent author; and no rational enquirer can, after serious reflection, suspend his belief a moment with regard to the primary principles of genuine Theism and Religion.

Excerpts from Section 15, general corollary, of his *Natural History of Religion* further reveal his views on God and nature where Hume's views are seen to not only be fairly traditional but also expressed with conviction and with, one might say, passion:

> Though the stupidity of men, barbarous and uninstructed, be so great, that they may not see a sovereign author in the more obvious works of nature, to which they are so much familiarized; yet it scarcely seems possible, that any one of good understanding should reject that idea, when once it is suggested to him. A purpose, an intention, a design is evident in every thing; and when our comprehension is so far enlarged as to contemplate the first rise of this visible system, we must adopt, with the strongest conviction, the idea of some intelligent cause or author. The uniform maxims, too, which prevail throughout the whole frame of the universe, naturally, if not necessarily, lead us to conceive this intelligence as single and undivided,

where the prejudices of education oppose not so reasonable a theory. Even the contrarieties of nature, by discovering themselves every where, become proofs of some consistent plan, and establish one single purpose or intention, however inexplicable and incomprehensible.

The universal propensity to believe in invisible, intelligent power, if not an original instinct, being at least a general attendant of human nature, may be considered as a kind of mark or stamp, which the divine workman has set upon his work; and nothing surely can more dignify mankind, than to be thus selected from all other parts of the creation, and to bear the image or impression of the universal Creator.

What a noble privilege is it of human reason to attain the knowledge of the supreme Being; and, from the visible works of nature, be enabled to infer so sublime a principle as its supreme Creator.

Thus, it is understandable why, in his closing remarks in *Dialogues,* Hume could not completely dismiss the teleological view of a designer behind nature. If we take this as the principle outcome of his life's work, we cannot wonder if in the end, that this was what he intended. Is it possible that, following the scholastic tradition, he made the strongest possible case for the teleological view, proceeded to tear it down, but then rather than triumphantly announcing the demise of the teleological view, he concluded that while his critique was legitimate and devastating as far as it went, there were *some things* that could not be taken down. The teleological view—perhaps beat up, bloody, and bruised—still stands, a monumental truth that cannot be rationally denied.

∾ ∾ ∾

While the traditional view throughout the nineteenth and twentieth centuries has been that Hume's assault did in fact demolish teleology, his arguments were posed within a set of naturalistic presuppositions that ignored the fundamental premises concerning the created order held by the earlier Christian natural philosophers. Hume argued from the mechanical view of nature. Present-day naturalistic philosophers now admit that his arguments were somewhat flawed. University of Wisconsin-Madison Philosopher Elliot Sober, in his book *Philosophy of Biology*[11] (1993) pointed out that:

> Although Hume's criticism is devastating if the design argument is an argument from analogy, I see no reason why the design argument should be construed in that way. Paley's argument about organisms stands on its own, regardless of whether watches and organisms happen to be similar. The point of talking about watches is to help the reader see that the argument about organisms is compelling.

Logical positivist and atheist Bertrand Russell also found the design inference to be a sound logical argument. He wrote in his 1945 book, *History of Western Philosophy*[12]:

This argument contends that, on a survey of the known world, we find things which cannot plausibly be explained as the product of blind natural forces, but are more reasonably to be regarded as evidence of a beneficent purpose. This argument has no formal logical defect; its premises are empirical and its conclusion professes to be reached in accordance with the usual canons of empirical inference. The question whether it is to be accepted or not turns, therefore, not on general metaphysical questions, but on comparatively detailed considerations.

~ ~ ~

As noted before, the "greater God" or "creation by mechanism" hypothesis was an important line of thinking that took shape in the eighteenth and nineteenth centuries in the move away from a "hands-on" God who intervenes directly in creation to a more detached Creator who uses "creation-by-mechanism" to accomplish his creative work. This thinking was reflected in the work of Kant. The idea of creation-by-mechanism is still prominent among many Christians today who hold the theistic evolutionary view.

Cornelius Hunter[13] calls this Kantian line of reasoning *theological naturalism* where it was held that a God who creates solely through natural laws is a greater God (the "greater God" hypothesis) than one who has to intervene in his creation to complete the job and make corrections along the way. Many theistic evolutionists today accept Darwinism on that basis but do so for *metaphysical* rather than *scientific* reasons. We note, however, that it is always true that in any reconciliation between Christian doctrine

158

and Darwinism, it is Christian doctrine that makes all of the concessions. Within the framework of this deductive form of reasoning the empirical evidence—or lack thereof—is pretty much irrelevant.

❧ ❧ ❧

When Darwin published his book *On the Origin of Species*, the arguments he set forth for his theory of evolution by natural selection was at best a speculative hypothesis with little direct evidence in its support, and he was forthright in admitting as much. However, that was good enough to serve the purposes of two important groups: the Christians who held to the "greater God" hypothesis and the naturalists who held to the "no God" hypothesis. Both were in need of a good mechanistic theory of life to support their respective metaphysical views, and that is what Darwin gave them. Well, sort of, but it was good enough.

Many embraced the "greater God" hypothesis. Unfortunately for some, they found that from this new vantage point of the world, the leap from "greater God" to "no God" was both compelling and "natural." After all, this was the way the winds of *progress* were blowing, and progress meant abandoning the old ways of thinking (religion and faith) for the enlightened ways of thinking (science and reason). Given the spirit of the times, "it took just a little nudge" (I think it was Benjamin Wiker who said this) "and that was all that was required to transform many a nominal theist into a budding agnostic. From there the gradual descent into full blown atheism began."

For many, that nudge was provided by nineteenth-century German theologian David Strauss (1808–1874) in his attacks on miracles, the historical authenticity of the Gospels, and the divinity of Christ. In the twentieth century, this became a well-traveled road for many who started their journey as Christians and, like Strauss, ended that journey as scientific materialist. Many today who call themselves "humanists" undoubtedly have their roots in this intellectual tradition.

It is deeply troubling that the speculations of two skeptics (Strauss and Darwin) in the nineteenth century, one a theologian and the other a naturalist, could have such a dramatic impact in the twentieth century in shaking the teleological foundations of Western civilization.

      ❦   ❦   ❦

The prevailing view in the nineteenth century was that sudden appearance of life (especially mankind) is an indicator of special creation. The only plausible naturalistic counterargument would be some sort of gradual cumulative process of change and progression from simple forms of life to complex forms of life. It was this notion of a slow cumulative history of life that Darwin had in mind as he thought about these things over the many years leading up to 1859. Further, these ideas had been "in the air" for several decades prior to Darwin, providing a basic framework of ideas that prepared the public for the reading of *Origins*. This prevailing atmosphere was heavily influenced by the anonymous publication in 1844 of *Vestiges of the Natural History of Creation*.[14]

Vestiges was the first attempt in intellectual history to connect the natural sciences with the history of creation and reflected the view that natural law, as a more exalted view of the works of God in creation, accounts for the history of life. *Vestiges* was controversial, widely read, and it became an international sensation, attracting the attention of the reading public as well as theologians, scientists, and even royalty. Prince Albert read *Vestiges* to Queen Victoria.

Vestiges argued for a general form of biological evolution (common ancestry descent with modification) but offered no mechanism to account for biological change and suggested a history of life that linked men to monkeys. It supported spontaneous generation of life and a continuing role for the Creator rather than pure "creation by law." In many ways, it was seriously flawed scientifically and underwent a series of revisions that continued to the publication of the twelfth and final edition in 1884 in which Robert Chambers, an intensely anti-evangelical middleclass Scotsman and political progressive, was revealed as the author. Chambers had died thirteen years before in 1871.

Vestiges was rejected (severely in some cases) by both clergy and scientists but praised by radicals and free thinkers. As for the public, opinions likely covered the entire spectrum, but probably what was most important was the fact of broad public interest. *Vestiges* was a great topic of discussion in Victorian England where talking about the latest ideas and trends was a popular pastime for the reading public.

In spite of its rejection by the clergy and science, *Vestiges* had a broad influence, resulting in a fundamental shift in thinking of the public and preparing the way fifteen years later for Darwin's *Origin of Species*. Without *Vestiges, Origins* would likely have had an even rougher time being accepted.

CHAPTER 10

COSMOLOGICAL DESIGN

In his paper[1] "The Place of Life and Man in Nature: Defending the Anthropocentric Thesis," Michael Denton introduced this topic with the following quote from Thomas Huxley[2]:

> The question of questions for mankind—the problem which underlies all others, and is more deeply interesting than any other—is the ascertainment of the place which mankind occupies in nature and of his relations to the universe of things.

Michael Denton also noted that the greatest discontinuity in nature is that which exists between the living world and the nonliving world. While the living world and the nonliving world are made of the same "stuff," the chasm between the two is vast, and there is no known naturalistic *bridge* connecting one to the other. There is nothing about the material world alone that gives even the slightest hint that something like life, consciousness, and mind should emerge from that world *on its own*. Mind may be *behind* the material world, but it is not *imminent in* the material world.

Within the naturalistic view of the world, these are great mysteries. Within the Christian view of the world, they are self-evident, essential aspects of the divine plan.

The view that mankind occupies a special place in the cosmos is found in the mythologies and creation stories of many cultures. This anthropocentric view of man has been especially strong in Western civilization due to the Old Testament account of creation and God's dealings with the Hebrews, the influence of Greek teleological thinking, and the Christian doctrines of the incarnation and redemption. However, with the advent 400 years ago of the scientific revolution, this anthropocentric view has slowly given way, at least in the academic world, to a rather mediocre view of man who, we are told, emerged from the natural world as an accident of nature. This view, however, is by no means unanimous, even among scientists. Supposedly, it was the hard, cold, uncaring facts of scientific discovery that led to this dismal view of mankind—but that is not true. It was the hard, cold, uncaring soul of *naturalism* that led us there.

But in spite of the *bad news* of naturalism, the *good news* of scientific discovery in the twentieth century is that the natural world points with overwhelming force to the centrality of mankind in the cosmic scheme of things where the significance of this discovery is in direct proportion to the prodigious size, splendor, beauty, and wonder of the cosmos.

Design in cosmology is defined in terms of fine tuning of the constants of nature and a long chain of very special but highly unlikely anthropic *coincidences*, all connected by the requirement that they support a certain "just-right" biochemistry on a "just-right" earth-like planet with a "just-right" amount of water and a "just-right" iron molten core gyroscopically stabilized by a "just-right" moon, all orbiting around a "just-right" sun that produces a "just-right" radiant environment necessary for the existence of advanced animal life and ultimately mankind in a universe violently hostile to life. But just as science (or rather, naturalism) tells us that there is nothing special about mankind, it also tells us that there is nothing special about our earth, our moon, our environment, our solar system, our galaxy, and our universe. Neither are we special. We're just very, very lucky. And to believe this, you have to either be very, very gullible or very, very "educated", or both.

✥   ✥   ✥

What is different about the science of our present times as opposed to the times of Copernicus, Galileo, Kepler, Bacon, Descartes, Newton, Faraday and Maxwell (to name a few) is its claim that the existence of mankind is simply a matter of chance and blind unguided natural processes alone. This modern version of the explanation for our existence can only be true, however, if the existence of the universe itself is a result of chance and blind unguided natural processes alone. If the universe was created by God, then anything that takes place in the universe is ultimately of divine origins. Scientists can claim that the origin of life can be explained solely through chance and blind unguided natural processes alone (and maybe someday it can), and they can claim that the emergence of mankind was through chance and blind unguided processes alone (even though they can't prove it), but if the universe itself was of divine origins, then life, consciousness, and mind are also of divine origins irrespective of the details of the mechanisms that brought them into existence or whether or not you call them "natural."

164

It is logically necessary that since the universe had a beginning, its existence is dependent on some *cause* beyond itself. To avoid an infinite regress, all such causal chains must eventually terminate in some self-existent *final cause* that of necessity is eternal, without beginning or end. In the West, we have traditionally identified this *final cause* as God, and this was the view of the early Christian natural philosophers. The only alternative to that view is that this *final cause* is material and mechanical and not divine, simply some kind of extension of the same mechanisms that account for the behavior and operation of the universe. However, such a universe would be both self-contained and self-existent, and that is generally held to be a logical impossibility.

Any quest for a *final cause* that has as its objective the elimination of God as a cause of existence will fail in that it will inevitably disintegrate into speculations about the unknown and the unknowable. Nevertheless, the *unknown* is a wonderful realm where chance, aided by imagination, can do wonderful things to sooth the souls of distraught naturalists who have to live in a real world that all day long, day in and day out, year in and year out, incessantly cries out "design and purpose." And it never shuts up! Never!

The only coherent view of a universe that includes a rational account of the existence of life, consciousness, and mind is that ultimately there is a single cause behind everything, and that cause must be self-existent and eternal, without beginning or end. Whatever that cause may be, it will leave its "fingerprints" *everywhere*. As stated in chapter 4, *the search for "fingerprints" and attempts to discern that to which they point is a perfectly legitimate activity of science*. But such an activity must be free to follow the fingerprints wherever they may lead.

❧ ❧ ❧

Charles Darwin postulated that life, through a long, gradual, cumulative process of small changes involving many generations, adapted to its immediate environment through the variation-selection mechanism (natural selection), and this is what accounts for the complexity of life, the design-like features we observe in living creatures and the emergence of mankind from lower forms. However, our modern understanding of life and its complexity, when considered in light of our modern understanding of the universe, suggests something radically different. It is no longer sufficient to consider (as did Darwin) simply the local environment and its effect on creatures who inhabited that environment as the means by which life evolved, eventually culminating in mankind. Now we must think the *other way around*. Forget about the influence of the environment as the dominant force in shaping life. There is little evidence that it did. Set aside the Darwinian notions of natural selection and the continuity of all life. Take a fresh look. Look at the basic facts. The universe came first, then life, then complex animal life, and finally, mankind. Mankind was the culmination of the creation process where the purpose of all that came before was to prepare a suitable home for him. There is far more evidence that suggests that the universe was specifically designed *for* man than that man was "designed" *by* the universe through natural unguided processes. This leads to a startling and profound proposition.

> *Life was not "fitted" to the environment. The environment was "fitted" to life long before life appeared.*

Accordingly, ideas about life and mankind were pre-existent to the universe itself. As Freeman Dyson[3] suggested in his article in *Scientific American* in 1971:

> As we look out into the universe and identify the many accidents of physics and astronomy that have worked together to our benefit, it almost seems as if the universe must in some sense have known we were coming.

All of the evidence says that Dyson was right. Admittedly, these observations are completely contrary to the neo-Darwinian narrative, but this is what the evidence from modern physics, biology, and cosmology tells us. Darwinism may have sounded like a great idea in Darwin's time but not in our time. Our modern understanding of science and reason simply rejects it.

◦§ ◦§ ◦§

Darwin's notion of *adaptation* (microevolution) as an explanation for *macroevolutionary* change was, in light of the very limited knowledge of both biology and cosmology that existed in late nineteenth century, a defensible hypothesis. However, in light of today's knowledge of the biological sciences, we now know that Darwin's adaptation cannot account for the macroevolutionary changes he postulated. Darwin should be recognized for the elegance of his theory of adaptation and the contributions he made to the biological sciences in his day. But at some point, we need to give up the misconceptions of the past along with our metaphysical prejudices and *accept the facts as they come to us from the natural world*. Supposedly, that is the way science is supposed to work.

Life may have been influenced to some extent by the environment, but overall body plans and fundamental features were not. Rather the environment was designed to provide both the raw materials from which life was formed and a suitable habitat in which life could exist, and there is no other way that this could occur except by design. These are the hard, cold, uncaring facts of scientific discovery, and they turn the naturalistic view of the relationship between the universe and mankind upside-down and inside-out. *Mankind is not an accident of nature. He is the purpose of nature.*

Alfred Russel Wallace, co-developer with Charles Darwin of the concept of evolution by natural selection (but who had much broader and deeper knowledge of the natural sciences than Darwin), anticipated these ideas in 1904 in his book *Man's Place in the Universe*[4]:

> Such a vast and complex universe as that which we know exists around us, may have been absolutely required . . . in order to produce a world that should be precisely adapted in every detail for the orderly development of life culminating in man.

In 1913, Lawrence Henderson, in his classic work, *The Fitness of the Environment*,[5] presented his defense of the notion of *the fitness of the environment for life* in opposition to Darwin's concept of *the fitness of the organism to the environment*. Darwin took a narrow view of the place of man in nature (an accident) and came up with evolution by natural selection without considering how the extraordinary conditions necessary for life arose. Henderson took a broader view and saw the logical necessity of a universe designed or "fitted" for life. Henderson's proposition was based on his observation of the facts about the environment and the understanding of life that were available to him at the time. Darwin's theory was based on a *hypothesis* that turned out to be contrary to the evidence yet is still accepted as fact.

One hundred years after publication of *The Fitness of the Environment*, advances in physics, cosmology, and biology have confirmed Wallace's speculations and Henderson's proposition, and soundly refuted Darwin's hypotheses. The environment within which life was to appear with all its bio-friendly properties came first and for all we know occurred in its fully completed form only one place in the universe—Planet Earth. There is no naturalistic account of how these properties arose. The unavoidable implication is that *the environment was designed for life*. Nature is not an accident, and if nature is not an accident, then neither are life, consciousness, and mind accidents.

Thus Henderson, in early twentieth century, "set a mark" before the natural sciences, warning of the possibility of a fundamental flaw in Darwin's thinking. That flaw was largely ignored until late twentieth century. Henderson's "mark" has emerged today in its fully realized form as cosmological design. If Henderson was right, then there is no theory of biological evolution. There was never a need for it. Darwin explicitly rejected design in nature and got it wrong. Henderson accepted design in nature and got it right. *Under Darwin there is no need for God. Under Henderson there is no need for Darwin.*

All of the evidence is leaning *toward* Wallace, Henderson, and God and *away* from Darwin. Nothing personal. I have absolutely no doubt that if Darwin were here today, he would enthusiastically agree. Such was the nature of his integrity as a scientist.

Perhaps it is time to retire Darwinism as a theory of the origin of species along with all of its materialistic accessories. It might, however, be retained in the curriculum of the biological sciences in public education as an interesting stage in the development of the life sciences and would be of particular value is showing students how metaphysical predispositions can influence science and absolutely corrupt science education.

✎ ✎ ✎

The general view today in the natural sciences is that if there is another planet somewhere in the universe (or maybe a planet in some other universe) that very closely matches the vast array of fine-tuned conditions and anthropic coincidences that pertain to *our* universe and *our* Earth, then carbon-based life will likely emerge on that planet and if it does, then through Darwinian processes, it will proliferate to fill, as chance may determine, the available life niches. Such is the power of natural selection and chance, which acting together, seem to take on the property of an irresistible driving force in biology just as gravity is an irresistible driving force in cosmology. If this is true, or if it is simply *held* to be true, then before that claim can be accepted as fact, there are two open questions that must be resolved with a convincing argument supported by *hard* evidence, not circumstantial fluff or speculative nonsense. The status of naturalism as a meaningful metaphysical alternative to theism rests on the validity of the answers to these two questions. Provide convincing naturalistic answers to these questions and naturalism may be accepted as the only rational philosophy of mankind.

The two questions: (1) How does science provide a naturalistic account for the origins of the universe and its "just-right" properties, and (2) how does science provide a naturalistic account for the origins of life, consciousness, and mind? If science cannot convincingly show that both can be answered in terms of unguided material causes alone, independent of naturalistic presuppositions, then naturalism has nothing to say about man because there is nothing about man that is naturalistic in the metaphysical sense of the word.

ఆ ఆ ఆ

It is refreshing to hear astronomers and astrophysicists, especially in the face of a prevailing naturalism in science, speak so openly about these amazing features, features that uniformly point to design and purpose and the centrality of man's place in the universe. You do not hear such *teleological* talk among biologists because in biology *you do not question Darwin*. Darwinism, naturalism, closed minds, and intolerance of other views seem to go together.

We look to scientists in the earth, planetary, and space sciences for their views on design in cosmology. The following quotes are from University of Washington scientists Peter Ward and Donald Brownlee taken from their book *Rare Earth*.[6] This book, written from the naturalistic point of view, is outstanding and highly recommended even by design proponents.

Most of the Universe is too cold, too hot, too dense, too vacuous, too dark, too bright, or not composed of the right elements to support life. . . . of all the known celestial bodies, Earth is unique in both its physical properties and its proven ability to sustain life. . . .

One of Earth's most basic life-supporting attributes is indeed its location, its seemingly ideal distance from the sun.

Without the moon it is . . . likely that no birds, redwoods, whales, trilobites, or other advanced life would ever have graced Earth. . . . Although there are dozens of moons in the solar system, the familiar ghostly white moon that illuminates our night sky is highly unusual, and its presence played a surprisingly important role.

Constancy of the tilt angle [of the earth's spin axis relative to its orbital plane around the Sun] is a factor that provides long-term stability of the Earth's temperature. If the polar tilt axis had undergone wide deviations from its present value, Earth's climate would have been much less hospitable.

If some god-like being could be given the opportunity to plan a sequence of events with the express goal of duplicating our "Garden of Eden," that power would face a formidable task. With the best intentions, but limited by natural laws and materials, it is unlikely that Earth could ever be truly replicated. Too many processes in its formation involved sheer luck.

From the biased viewpoint of Earthlings, however, it does appear that Earth is quite a charmed planet.

Many other scientists agree. Michael Denton said in Nature's Destiny[7]:

It is as if from the very moment of creation the biochemistry of life was already preordained in the atom-building process, as if Nature were biased to this end from the beginning.

In his book, *God and the Astronomers,*[8] NASA astronomer Robert Jastrow said:

For the scientist who has lived by his faith in the power of reason, the story ends like a bad dream. He has scaled the mountains of ignorance; he is about to conquer the highest peak; as he pulls himself over the final rock, he is greeted by a band of theologians who have been sitting there for centuries.

Amherst astronomy professor George Greenstein in his 1988 book *Symbiotic Universe: Life and Mind in the Cosmos*[9] said:

As we survey all the evidence, the thought insistently arises that some supernatural agency—or rather Agency—must be involved. Is it possible that suddenly, without intending to,

173

we have stumbled upon scientific proof of the existence of a Supreme Being? Was it God who so providentially crafted the cosmos for our benefit?

NASA astronomer John O'Keefe said[10]:

If the Universe had not been made with the most exacting precision we could never have come into existence. It is my view that these circumstances indicate that the Universe was created for man to live in.

Nobel Laurate Arno Penzias said[11]:

Astronomy leads us to a unique event, a universe which was created out of nothing and delicately balanced to provide exactly the conditions required to support life. In the absence of an absurdly-improbable accident, the observations of modern science seem to suggest an underlying, one might say, super-natural plan.

British astrophysics Edward R. Harrison, in his book *Masks of the Universe*,[12] first published in 1985, said:

Here is the cosmological proof of the existence of God— the design argument of Paley—updated and refurbished. The fine tuning of the universe provides prima facie evidence of deistic design. Take your choice: blind chance that requires

multitudes of universes, or design that requires only one. . . .
Many scientists, when they admit their views, incline toward
the teleological or design argument.

British astronomer, mathematician, atheist and relentless critic of
Darwinism Sir Fred Hoyle said[13]:

> A common sense interpretation of the facts suggests that
> a superintellect has monkeyed with physics, as well as with
> chemistry and biology, and that there are no blind forces worth
> speaking about in nature. The numbers one calculates from the
> facts seem to me so overwhelming as to put this conclusion
> almost beyond question.

A quick check will show that with the exception of Denton, none of
these scientists are biologists. Just as Denton noted that there is a vast
chasm between the living and nonliving worlds, there also seems to be
a vast chasm between the life sciences and the physical sciences. Many
physicists and astronomers seem to be comfortable with the teleological
view of the world. Not so for most biologists.

❧ ❧ ❧

If we limit ourselves to a consideration of the only universe we know
anything about (sample space of one), then it is difficult to escape the
anthropocentric message it contains. Escape is not so difficult, however,
if the question is taken out of the context of our one and only universe

(which we know quite a bit about) into a much broader context of a near infinitude of *unobservable* universes (which we know nothing about and can never know anything about and which probably exist only in our imagination). But if we assume randomly varying properties in this infinitude of universes, then somewhere in this maze of universes (that we cannot see and which we can know nothing about and which probably do not exist), there must be one (even though it probably does not exist) that matches the properties of our own bio-friendly universe so completely and to such a high degree of precision that it is inevitable that life will emerge somewhere in that universe after which natural selection will take over in filling the evolutionary "niches" with life that of necessity will look very much like life on earth, and with a little more luck (which is never in short supply for the naturalistic mind), something like man will eventually evolve complete with consciousness and mind and who most probably will speak English with a British accent and must have his tea and biscuits at exactly 3:10 in the afternoon. And there also will be another universe that is exactly the same but where they have their tea and biscuits at exactly 3:20 in the afternoon. And then there must be—blah, blah, blah. Within the *multiverse hypothesis*, anything and everything is possible—except for one thing. There can be no God.

The whole point of *multiverse* is to show how it is "reasonable" to believe that our own universe with its special properties necessary to support intelligent life is simply *the one* universe in an infinity of universes that, without knowing or caring, won the lottery. There is simply no need to look to a transcendent intelligent designer to account for our special universe and life. There really is no design in biology; it just appears that way. The universe really isn't fine-tuned for life; it just appears that way.

If it did not appear that way, we would not be here to observe that it did not appear that way, but here we are and, sure enough, it appears that way, and therefore naturalism is true even though if it were true, we would not be here to observe that it did not appear that way but—and around and around we go.

In spite of the absurdity of the multiverse hypothesis and in spite of the fact that it does not give an account of how our present universe came into existence (much less how an infinity of universes came into existence), there is the possibility that in time it will, of metaphysical necessity, become the *consensus view* within the scientific establishment concerning the cosmos, just as neo-Darwinism became the consensus view within the scientific establishment concerning life. The naturalistic view of the world demands it, and methodological naturalism provides the protective services.

❧ ❧ ❧

At the heart of the multiverse argument is the same logic that under-lies the *infinite monkey theorem*[14] that says that given enough monkeys pounding away on typewriters and given enough time, they will eventually produce the works of Shakespeare. As with any probabilistic analysis, certain assumptions must be made. In the infinite monkey theorem, we must assume an unlimited supply of typewriters and monkeys and bananas, the monkeys will actually hit the keys rather than just defecate on them (as empirically demonstrated), and the keys they hit will be selected randomly, which turned out to be a really bad assumption. For some reason they seemed to like the letter "s". (Maybe they knew they were *S*imians).

The monkey experiment did nothing to give the impression that "given enough time, anything can happen." Of course, no one ever thought it would. Even Darwinists need to take a little time off for fun.

This amazing creative power of chance and time was touted in 1954 by evolutionary biologist George Wald in an article[15] published in *Scientific American* "The Origin of Life" (Wald was later to receive the Nobel Prize):

> Given enough time it ["it," meaning the origin of life] will almost certainly happen at least once . . . Time is in fact the hero of the plot . . . Given so much time, the 'impossible' becomes possible, the possible probable, and the probable virtually certain. One has only to wait; time itself performs the miracles.[16]

But Professor Wald is in error. To say that *anything* can happen is equivalent to saying that *everything* can happen. Given the physical laws and principles of nature, there are some things that are not possible, and therefore *anything* cannot happen. The Second Law of Thermodynamics tells us that eventually hell will freeze over, but *anything* cannot happen before it freezes over and probably nothing interesting happens after it freezes over. Even Lewis Carroll's character *Alice* understood this (Michael Denton made this exceedingly appropriate observation in his book *Evolution: a theory in crisis*[17]):

> Alice laughed. "There's no use trying," she said: "one *can't* believe impossible things." "I daresay you haven't had much practice," said the Queen. "When I was your age, I always did it

for half-an-hour a day. Why, sometimes I've believed as many

as six impossible things before breakfast."

As for Professor Wald, in the 1980s, he underwent an astonishing transformation[18] in his thinking from the atheism reflected in the *Scientific American* article to what might be described minimally as *scientific deism*, and it was the anthropocentric nature of the cosmos and the problem of providing a naturalistic explanation for consciousness and mind that inspired this transformation. Interestingly, Wald came to the conclusion that of necessity *mind had* to come *before matter*, and in fact, *mind* has always been "there." *Mind* is eternal, and it was this eternal *mind* that created the universe and life. That is the only rational account that can be given for our existence. Wald also noted that physicists did not seem to be all that troubled by his thinking, but not so with his fellow biologists.

You might say that Professor Wald saw the *light*, the light of reason. Good for you, Professor Wald! Good for you!

The idea that anything can happen runs head-on into common sense. For a natural system, that is, a system within the natural world (or the natural world itself), what can and cannot happen is determined by the laws and principles that regulate that system. Chance applies to such systems but only within the range of outcomes permitted by the laws and principles of that system. Likewise *processes* that take place within such a system are also bound by the laws and principles of that system and have no powers beyond those which already reside within that system. It is absurd to say that within such a system "anything" can happen. Chance has no creative power. Naturalistic scientists just say that it does. The Red Queen would probably agree. If you think like the Red Queen, anything can happen, even impossible things. You just have to believe it, and it will be true, even before breakfast.

The Red Queen was a positivist, who, unhindered by reality, was free to create her own version of reality and because she was the Queen, it was the law of the land and you best agree with her or, "off with your head". But, if you think scientifically, like Alice, who was a realist, you accept that there are things that cannot happen as a matter of principle. You can wait as long as you like but, if it can't happen, it won't happen.

You may see some similarities here between what goes on in Darwin-land and what goes on in Wonderland where, if you disagree with their rulers, both enforce an "off with their heads" policy.

Outcomes other than those permitted by the system and its laws can, however, occur due to influences external to that system and can do so without violating the laws of the system. Take a modern jetliner, such as the Boeing 787 *Dreamliner*. The 787 is an example of how intelligent agents use their knowledge of the "system and its laws" to create technological wonders for a purpose determined beforehand which natural systems, left to themselves, can never create. In the process of designing, manufacturing, and operating a Boeing 787, no physical laws or principles are broken. The entire enterprise rests on the ability of intelligent agents to *unify physical laws with design principles around a common purpose*: the design, manufacture and operation of the 787.

The universe in which we live is the only one we know anything about and the only one we *can* know anything about. There is nothing fundamentally wrong with speculations that are related to some observable feature of the universe but speculations about imagined universes—billions and billions of them? Such excursions into science fiction are motivated, not by physical evidence of the existence of billions of other universes, but by naturalistic dogma.

> *Multiverse is a deduction from metaphysical necessity, not a scientific finding based on observation, experimentation, and inductive reasoning.*

Consequently, as Paul Davies points out in his book *The Goldilocks Enigma*,[19] multiverse is not widely embraced, and some scientists even refer to it as pseudoscience. Most would say that if you cannot conceive of a way in which to test a hypothesis, then it cannot be considered a true scientific hypothesis. Or, they would just say it is not "science." Multiverse is clearly in that category.

❧ ❧ ❧

At the turn of the twentieth century, science had a very limited understanding of the size and structure of the universe, and most astronomers assumed, along with Aristotle, that the universe was eternal, without beginning and without end. (This assumption was also favorable because it solved the *contingency* problem.) The best we could tell from the astronomy of that day was that the Milky Way galaxy established the physical bounds of the universe. Robert Jastrow, in his book *God and the Astronomers*[20] (1978), tells the fascinating story of how astronomers and physicists in early twentieth century began to put together the pieces of a puzzle in which a picture of the universe began to emerge. The picture that emerged was not of a static and eternal universe as most scientists at the time had assumed, but of a dynamic universe that was enormous in size, very old, increasing in size at high speed and most startling of all, had a beginning. Jastrow called it the scientific story of Genesis:

> Now we see how the astronomical evidence supports the biblical view of the origin of the world. The details differ, but the essential elements in the astronomical and biblical accounts of Genesis are the same: the chain of events leading to man commenced suddenly and sharply at a definite moment in time, in a flash of light and energy.

We review this story as recorded by Jastrow because it provides an important account of how scientists, in spite of their aversion to the idea of a beginning—especially one that looks a lot like Genesis—were eventually willing to follow the evidence in spite of the threat it posed to their metaphysical predispositions. One could hardly imagine a more controversial subject for a "shoot-out" between science and religion than Genesis 1:1—or, as it turned out in this case, a grand reconciliation.

The first hint of the coming revolution in astronomy was the observation in 1913 by Vesto Melvin Slipher of about a dozen galaxies moving away from the earth at high speed as evidenced by the red shift of their electromagnetic spectrums. Astronomers at the time did not know what that meant, but they sensed that it was of great importance. Then in 1917, Einstein published his equations on the general theory of relativity but failed to make the connection between his theoretical work and cosmology. The Dutch astronomer Willem de Sitter almost immediately discovered a solution to Einstein's equations that predicted an expanding universe. Then in 1922, Russian mathematician Alexander Friedmann found a subtle but important algebraic error in Einstein's equations (division by zero, a serious "no-no" in mathematics), thereby revealing another solution to Einstein's equations for an expanding universe. Einstein, in 1923, reluctantly acknowledged his error. Friedmann's interests were limited to abstract mathematics where he never viewed his theoretical work as actually having any physical meaning. Nevertheless, by this time, the picture of a universe with a beginning had emerged in the minds of the astronomers and with it, signs of a growing irritation among scientists, especially Einstein. They were deeply troubled by the idea of a beginning and its metaphysical implications.

By 1925, more galaxies were observed, which were receding away from the earth at high speed, further confirming the notion of an expanding universe. Astronomers Edwin Hubble and Milton Humason using the 100-inch telescope at Mt. Wilson measured not only speeds but distances of the receding galaxies. A three-dimensional picture of the universe was beginning to take shape in which the Milky Way galaxy was just a tiny island in a vast sea of such islands. From this data, Hubble also derived his famous constant, relating the speed of receding galaxies to their distance from the earth.

In 1927 the Belgian priest and physicist Georges Lemaitre (PhD in physics, MIT) who had physical reality rather than mathematical abstraction in mind, derived his model of an expanding universe from Einstein's equations, the same solution as derived by Friedmann in 1922 of which he was unaware (no Internet at the time), and it was the combination of his theoretical work and the experimental work of Hubble that kicked off the revolution in astronomy. Lemaitre reasoned that at the beginning, all space-time-matter was contained in what he called the "primeval atom" from which the universe emerged in an explosive event, which Fred Hoyle many years later called the "Big Bang."

During this period, astronomers were also beginning to recognize that the Second Law of Thermodynamics, just like Newtonian mechanics, must apply to the cosmos as well as on earth, and that meant that the universe must be running down, proceeding with time from a state of high order (low entropy) to a state of ever-decreasing order (increasing entropy), further discounting the idea of an eternal universe and supporting the idea of a universe with a beginning. This observation introduced another puzzling and fundamental question. How did the universe acquire its low-entropy state early in its history? In 1955, shortly before his death, Einstein finally accepted that the universe had a beginning.

Finally, the accidental discovery in 1965 by AT&T scientists Arno Penzias and Robert Wilson of the 3 K cosmic microwave background radiation (later refined to 2.7 K) predicted by Big Bang theory, settled the matter for most astronomers, securing the position of Big Bang cosmology (despite its powerful religious implications) as the accepted scientific theory for the origin and history of the universe.

To be clear, the final chapter on Big Bang cosmology has not been written. In fact Big Bang cosmology has always had "problems," but it has always been assumed (at least by most) that eventually further discoveries would finally make sense of it all, but that—at least so far—is not the way it progressed at all. In fact, it has gotten (in nonscientific terms) downright "crazy." In the early 1990s, astrophysicists found evidence for the existence of what came to be called "dark matter." Distant galaxies were observed that are *accelerating* due to some mysterious energy source, which was aptly given the name "dark energy." Given the equivalence of mass and energy in physics, it has been determined that together, dark matter and dark energy comprise some 95 percent of the mass of the universe. Other distant galaxies were observed that appear to be older than the universe itself. In light of these and other developments, you can hardly claim with certainty that Big Bang cosmology is a correct theory. The difference is, unlike biologists, cosmologists openly acknowledge the problems with the Big Bang model.

Critique of Big Bang theory should not be viewed as a criticism in general of the scientists who built that theory. Rather it is a testimony to their diligence and ingenuity as scientists. They followed the evidence, and this is where it led them. They are still following the evidence. That is why we have problems. These "problems" simply tell us that science is standing at the threshold of some dramatic cosmological discoveries which, when you look at the big picture, portends new discoveries in physics as well. Why so? Because the mechanics of the universe is due to physics, and if there are things observed in the cosmos that cannot be understood in terms of physics as we understand it today, then perhaps the secret to that understanding will be found in our understanding of physics, not in the cosmos.

Lemaitre set a high standard for how science should deal with metaphysics. Separate the science from metaphysics of any kind. How do you do that? Follow the evidence wherever it may lead in spite of where it may lead. Science should simply ignore whatever religious implications might emerge, and let the facts fall where they may. What is important is to perform scientific work, based on observable physical evidence and do so with objectivity and integrity. The philosophers and theologians can then sort out the metaphysical implications.

Thomas Huxley had good advice[21] for all when he said:

> Science seems to me to teach in the highest and strongest manner the great truth which is embodied in the Christian conception of entire surrender to the will of God. Sit down before fact as a little child, be prepared to give up every preconceived notion, follow humbly wherever and to whatever abysses nature leads, or you shall learn nothing.

❧ ❧ ❧

One can hardly imagine any event more thoroughly imbued with religious significance than the creation of the universe because all that follows that creation event will be unavoidably and permanently "marked" by the ontological principles that governed that event. These markings are the so-called "fingerprints" of creation. Every aspect of the natural world is so marked. In early twentieth century, science came face-to-face with these markings in the cosmos, and despite the metaphysical discomfort experienced by most of the scientists involved (Lemaitre was a notable exception), the force of the theoretical and empirical evidence eventually prevailed. By their example, these scientists, painful as the process was for some, revealed an underlying principle of science that has been largely abandoned. If you are going to do scientific work, you must follow the evidence wherever it may lead, regardless of its metaphysical implications. Physical evidence reliably obtained cannot be discarded because scientists are opposed to the marking it received at creation. To do so is to largely abandon scientific methodology and completely abandon intellectual integrity. Such are the fruits of methodological naturalism.

Early twentieth-century scientists, however reluctantly, remained faithful to this principle in spite of their metaphysical preferences. They intuitively knew that as scientists they could not do otherwise. They followed the evidence, they accepted the markings, and it led them to an astounding understanding of the universe—and that journey is hardly over. The trail is heating up. There are exciting times ahead.

CHAPTER 11

UNIFICATION

Physicists dream of unification. From fundamental particles to galaxies, physicists intuitively *know* that somehow the laws of physics must be unified into a single coherent and complete system of laws, equations, principles, and mechanisms that explain all physical phenomena. Unification is the holy grail of physics, and physicists are confident that when it all finally comes together, it will be beautiful. Quantum mechanics and gravity will be reconciled, we will finally understand dark matter and dark energy, and the final chapter will be written in securing scientific materialism as the only reliable basis for understanding ourselves and our world.

There is every indication however, that this is not at all how this great human drama will end. Read on.

❧ ❧ ❧

Earlier, in chapter 6, we posed a rational metaphysically-neutral meaning of *natural causes*. We said that "natural causes are those causes

188

that are present and operative in the natural world that produce observable effects that can be subjected to scientific study." Through the process of observation, experimentation, and inductive reasoning, we infer laws, such as Newton's laws of gravity that describe planetary motion, Maxwell's equations that describe electromagnetic phenomena, and Einstein's space-time physics. If we accept that the meaning of *physical law* incorporates both law and the mechanism it regulates, then we can speak generally of physical law as a *cause* in nature. In this sense then, physical law qualifies as a *natural cause* because it is present and operative in the natural world, and it produces observable effects that can be subjected to scientific study. Its primary function within the natural sciences is not in what it *causes*, but in what it *explains* and in *how* it explains it. Consequently, we take as a given that physical *law* is a *real* part of the natural world, a natural cause, and thus subject to scientific investigation. This says nothing, however, about the ultimate origin and cause of these physical laws or why they take on the mathematical form they take on or serve the function they serve. And it does not even begin to explain the extraordinary order we observe in nature, even though these physical laws are very much an essential part of that order. *Laws* are what they are because *order* is what it is and *order* is what it is because *laws* are what they are. Law and order are inseparable in both a well-ordered society and in a well-ordered nature.

The ultimate origin and cause of the physical laws we see at work in the natural world are generally of little concern to the normal practice of the scientist and more a subject of interest to philosophers and theologians. But all — scientist, philosopher, and theologian — should be interested in the fact that the existence of physical laws infers the existence of a *law-giver*. Law is an abstract conception of mind. Somewhere there has to be a mind involved in that conception. How do you get a *law* without a *law-giver*, and where do you find a law-giver who has the power to enact laws on such a grand scale as that of gravity, electromagnetism, and space-time? *That* law-giver can only be *The* Law-Giver, God Himself. Newton and Maxwell, as devout Christians, clearly understood this. Modern naturalistic scientists, however, reject the idea of a divine law-giver for purely metaphysical reasons while accepting the place of physical law in the natural world and thus in the natural sciences, quietly assuming that since physical law is a *natural* phenomenon, then its ultimate origins *must* also be natural. But such reasoning is non sequitur. It is not necessarily true.

But *design* is also an abstract conception of mind, and if we were to treat *design* just like we treat *law*, as a *real part* of the natural world that is present and operative within the natural world, and that produces observable effects that can be subjected to scientific investigation, what might we expect? Are we justified in even proposing such a thing? While this is a horrifying thought to naturalists, is there some fundamental reason (other than their metaphysical horror) why we can't do this simply as an exercise in reason? We accept that this way thinking makes sense for the concept of *law*, so why not *design?*

The scientific establishment would probably say that trying to blend teleological and scientific ideas is nonsense, like blending science and religion. Further investigation, however, shows that *law* and *design* in fact share important ontological properties. Both point to intelligent agency, both are abstract conceptions of the mind, they are of the same logical category, the effects of both are observable, and consequently both may be subject to inductive reasoning. Consequently, it is reasonable to suppose that they should be treated the same way within the natural sciences and that means they should be treated the same way under methodological naturalism. But they are not, and this is an important question that demands an answer whether put to a high school biology teacher, a local school board president, a state legislator or the National Academy of Sciences. I, for one, would be very interested in the answer.

The similarities between law and design are easily demonstrated. Consider the genetic code which is central to cellular operations and consequently life itself. Most would agree that the genetic code exhibits design-like features pointing to the work of a designer, an intelligent cause. Is the genetic code present and operative in the biological cell? Yes. Have we deciphered the code and the way it is impressed on matter? Yes. Can we observe its effects? Yes. Do we understand its role in cellular operations? Yes. Can we subject it to scientific investigation? Yes. Does such knowledge advance the biological sciences? Yes. Is it scientific? Yes. Has science conclusively and rigorously determined that natural law and chance processes alone not only *can,* but *do* account for the genetic code and its function in biology? No.

Let us now consider the concept of the law of gravity and its role in planetary motion to the same regimen of questions. Most would agree that the law of gravity exhibits law-like features pointing to the work of a law-giver, an intelligent cause. Is the law of gravity present and operative in the natural world? Yes. Have we discovered this universal law of gravity and its influence in the natural world? Yes. Can we observe its effects? Yes. Do we understand its role in explaining planetary motion? Yes. Can we subject it to scientific investigation? Yes. Does such knowledge advance the physical sciences? Yes. Is it scientific? Yes. Can we account for the existence of the law of gravity through natural, unguided material processes alone? No.

One final question concerning both: Do the existence of *law* and *design* in nature have religious implications? Yes, but those implications are ignored for the law of gravity.

It would seem that because of its implications for a *law-giver* behind the natural world, perhaps methodological naturalism should step in and declare that the concept of law is religious, not scientific in nature and put a stop to this creationist nonsense. But we hear not a peep out of naturalistic scientists and understandably so since science without physical law is not very interesting, and we would find ourselves once again wandering in the wilderness of Aristotelian natural philosophy. So why does *law* get a pass from methodological naturalism, but *design* is banished to the outer darkness?

My assessment is that the concept of *law* gets a *pass* from methodological naturalism because of its historical association with the *mechanical philosophy* that emerged in the seventeenth century from Newton's laws of mechanics and was embraced by naturalists from the eighteenth century to our own times. Thus, historically, the concept of design is associated with the teleological view of nature while law is associated with the mechanical view of nature. Mechanics is *natural*, design is *religious*, and it is this artificial convention founded not on reason but on historical contingency that separates them in science in spite of the fact that they are in fact inseparable in nature. Both require intelligence, but more than that, these two aspects of the natural world must be *unified*. Thus, with respect to both scientific and metaphysical principle, law and design are inseparable partners in the natural world.

So where is the conflict? There is no conflict, at least not between science and religion; rather, the conflict is between two religious views of the world: the naturalistic view and the teleological view.

This discussion is well on the way to its "natural" conclusion. But before we continue, you should be aware that to proceed might be somewhat like taking a bite of forbidden fruit, metaphysically speaking, which introduces you to some strange and wonderful "flavors" you never encountered before. If you take that bite, your world might be forever changed—maybe even for the good—but you won't know unless you take the bite. For the naturalist who does not want to be troubled by other philosophies, it might be best to not take the bite. Just stick you head back into the sand. But how can you live without knowing? Go ahead. Take the bite. Eve did.

Design *must* be accepted as a *natural cause* for exactly the same reason that physical law is accepted as a *natural cause*. Why? Because of its explanatory power. The effects of both *law* and *design* are present and operative in the natural world, and we know this is true because we can observe their effects. If we did not see their effects, we would have no reason to relate either *law* or *design* to the natural world and its operation and behavior. In one case we infer *law*. In the other case, we infer *design*. In fact, law and design must be fine-tuned to "fit" each other to serve some intended end. We detect both *law* and *design* through observation, experimentation, and inductive reasoning, and both point to intelligent agency. The justification for treating *design* as a *real part* of the natural world is exactly the same as it is for treating *law* as a *real part* of the natural world. You cannot explain the natural world apart from either *law* or *design*. You may argue all you want that both are the result of unguided chance processes, but if you cannot account for those processes that supposedly produced those laws and designs (not to mention the existence and properties of the material world in which they operate), your explanation is empty. Further, whatever religious status you assign to *design*, you must assign to *law*. Logically, you cannot have it both ways. You cannot say that *law* is *science*, and *design* is *religion*. Consistency demands that you have it one way or the other. Reason tells you which way is the right way.

∽ ∽ ∽

Both *law* and *design* are present and operative in nature, and both are nonphysical conceptions of the mind that are unobservable except through their effects. Law points to a law-giver. Design points to a designer. Just as we determine that *law* is present and operative in nature by inference from its effects, so also we determine that *design* is present and operative in nature by inference from its effects.

If you can observe it, it is natural, and if it is natural, it has a natural cause, and whether that cause reflects law or whether it reflects design, it can be subjected to scientific investigation.

Henceforth, I will capitalize *Law* and *Design* to signify the significance of their place in the natural order and thus, in the natural sciences. As for *chance*, it may be capitalized in Las Vegas, but not in the natural order and not here.

194

What then is the proper relationship between Law and Design? Is there anything about the natural order that connects the two, a *necessary* connection that accounts for the unique natural order we observe?

The order and rationality we observe throughout nature lead us to an astounding proposition: Law and Design are unified in nature, and if they are unified in nature, then there must be some principle around which they are unified. If there is such a principle that does in fact serve that unifying function, then this would be a direct indication of intelligent agency and *purposeful* design where the equations of physics and the properties of matter, energy, and space-time and the ordering of every aspect of the universe are "fitted" *to each other* to satisfy this unifying principle set down according to a master plan from before the beginning of time to accomplish some predetermined universal outcome. Scientific discovery has revealed that every aspect of both the living world and the nonliving world appear to be "fitted" according to a single unifying principle: *the centrality of mankind's place in the cosmos*. Law and Design are unified in nature around the centrality of the place of man in the cosmos. If this were not so, there might still be some sort of physical reality, but there would be no one to observe it or to ask such questions and form such conclusions.

But, there is a "loose end" to this rosy picture of unification. We are still faced with the fundamental problem in physics of the unification of quantum mechanics and general relativity. An all-wise and all-powerful creator who is supposedly the author of all that is good, true and beautiful would never create such a monstrosity. He didn't. The flaw is in our understanding. Let not your heart be troubled. In due time, physicists will eventually resolve this matter and when they do, it will, as they have always said, "be beautiful". The rational design of the universe and the searching heart of man absolutely guarantee it!

❧ ❧ ❧

The discovery of purpose in the universe by means of the natural sciences is a stunning development in the history of mankind (appalling to some). We have always suspected that there was something special about the world that we could not fully grasp on our own (we all seem to have something like Psalm 19 written on our hearts). But through the natural sciences we have discovered that the natural world, as announced in Psalm 78:2, is now "uttering hidden things from of old," things that are comprehensible to any who will listen. Discoveries in the last 100 years in the cosmological and biological sciences reveal the unification of Law and Design in nature around the centrality of man's place in the universe.[1] This is the message of the cosmos—*the Gospel of the Cosmos*.

Naturalistic science and the consensus view of the scientific priesthood notwithstanding, there is substantial evidence that supports the thesis that the two great explanatory principles of the natural world are not *Law* and *chance,* but *Law* and *Design*, and that they are unified in the natural world. In this causal system, *chance* is still important, but it is axiomatic that in a rational universe *chance* must forever be confined to the limitations imposed by the laws and principles of the system within which it operates. No rational argument or experiment has ever proven otherwise. Evolution is such an experiment, and we are still waiting—and waiting—and waiting. Time ran out on Darwin's experiment long ago.

Ultimately, science that is rooted in a naturalistic commitment is irrational because through that commitment, it has rejected an essential feature of reality: the rationality of the created order. Compelled by its naturalistic commitments and its deep loathing of design, the scientific establishment has constructed a framework of concepts based on chance and necessity alone, despite the fact that these concepts have not been able to provide a plausible account for the origins of life, the origins of man, the emergence of consciousness and mind, or the anthropocentric nature of the universe. To attribute all of this to chance processes is utter and complete nonsense. Law and Design unified around a single principle can do these things, but it takes the supernatural to pull it off. This is where all the evidence points.

❧   ❧   ❧

One of the most profound questions philosophers have asked throughout history is "What is the relationship between mankind and the universe?" Are we accidents of nature, meaningless aberrations on a speck of dust adrift in a vast, uncaring universe, or is there some deep transcendent meaning and purpose to our existence? Are we in some profound way connected to the universe?

The so-called *Copernican principle* says that there is nothing special about man because there is nothing special about his place in the universe. This so-called "principle" was framed as an answer to this question based on the proposition as set forth by Copernicus that the sun, not the earth, was at the center of the universe. Much significance has been attached to this little wisp of cosmological trivia over the years, even though anyone who knows anything about Christian doctrine and Aristotelian natural philosophy may wonder what all the fuss is about. There is nothing in Christian doctrine that in any way relates man's physical location[2] in the universe to his status in the eyes of God. After all it is the creation, the incarnation, redemption and, in our own time, our understanding of the centrality of the place of man in the design of the universe that reflect the unspeakable favor shown by God to man, not his location. *Status* due to *location* is an Aristotelian concept except that in the Aristotelian world, if you are at the center of the cosmos, you live in a cesspool of corruption—not all that special after all. In the Aristotelian system, perfection was found only in the celestial spheres.

While the Copernican principle may have been devised as an attempt to belittle man's place in the universe (at least in the eyes of men), even the most egocentric among us could not have imagined a more exalted status of man than that revealed by the natural world itself. Even though the scientific priesthood claims that man is an accident of nature, what has emerged from scientific discovery in the twentieth century is a radically different message, a message rooted in biology, physics, and cosmology. It is a personal message, a message for you and for me, for the world, for the naturalist and the believer alike. We are not *accidents* of nature; we are the *purpose* of nature.

Through the natural sciences we have "read" the *book of nature*, and in that reading, we have found the "fingerprints" of the Creator throughout all of creation from the fundamental machinery of life to the architecture of the universe. We have found the *bridge* that connects the living world to the nonliving world, and it is not the bridge the naturalists have so hoped for.

◈ ◈ ◈

Michael Denton, in the prologue of his *book Nature's Destiny*,[3] contrasts the present materialistic and secular view of the world held by the academy with that held by the natural philosophers of Europe in late fifteenth century shortly before the birth of modern science. Here is Denton:

> For Christian scholars, the biblical revelation, and particularly the Incarnation, sanctioned the profoundly anthropocentric character of their medieval worldview. The extraordinary anthropocentricity of the culture of the Christian Middle Ages was wonderfully conveyed by Aron Gurevich in his classic work *Categories of Medieval Culture*[4] . . .

And here is Gurevich:

> The effort to grasp the world as a single unified whole runs through all the medieval summae, the encyclopaedias and the etymologies. . . . The philosophers of the twelfth century speak of the necessity of studying nature; for in the cognition of nature in all her depths, man finds himself . . . underlying these arguments and images is a confident belief in the unity and beauty of the world, and also the conviction that *the central place in the world which God has created belongs to man*. (my emphasis)

The views of the medieval philosophers were in a sense prophetic for they have been fulfilled in our own times, not in terms of the opinions of modern-day theologians and biblical scholars, but through the natural sciences and *that* despite the skepticism that rules the academy.

CHAPTER 12

PHYSICS DERAILED

I f we took a view of physics from the perspective of late nineteenth century, we could say with some confidence that the spectacular success of the natural sciences has affirmed the credibility of the fundamental presuppositions and methods employed by the founders of modern science as a means for gaining reliable knowledge about the workings and operation of the natural world. As described in chapter 3, we can attribute this success largely to the work of Christian natural philosophers whose view of the world as a created order provided the unique mindset that made science possible. They were realists, and in the early decades of the twentieth century the mechanics of Newton and the electromagnetism of Maxwell, together known as classical physics[1], along with the space-time physics of Einstein, provided what appeared to be a complete and comprehensive foundation of physical laws, principles, and mechanisms that could, with the exception of life, explain all physical phenomena.

These were the fruits of 400 years of science undergirded by the realist epistemology. But, in late nineteenth century, signs of trouble began to appear.

In late nineteenth and early twentieth century, physicists were attempting to make sense of the atomic world in terms of the laws and principles of classical physics, but they found that certain phenomena such as blackbody radiation and the photoelectric effect made no sense within the classical paradigm. Further, it seemed that at the scale of the atom, energy was not continuous but rather was quantized and that classical physics (at least as they understood it at the time) simply did not "work" at the microscopic level of the atom. Given these difficulties regarding classical physics and the atomic world, these supposed limitations of classical physics seemed reasonable opening the door to a new epistemology called positivism that would hopefully permit atomic physics to progress beyond its stalemate.

From our[2] early twenty-first century perspective, it is difficult to comprehend the idea that there might be some fundamental problem with quantum physics given the spectacular technological advances of the twentieth century based on that physics that have transformed our lives and continue to do so with no end in sight. How could so much have been accomplished unless quantum physics was basically correct? But after more than half a century, there is little that can be said regarding progress in physics itself. Thanks to the technological advances, we have a broader and deeper understanding of the natural world than we did fifty years ago, but that understanding has also introduced baffling unexplainable mind-boggling paradoxes. We encounter these paradoxes, not at the periphery of physics, but at its core. Something is fundamentally wrong with our present conception of atomic physics, and there is little indication that under the present paradigm that anything will ever change. Sometime around the 1960s, physics got "stuck in time" in some kind of squirrely wonderland, about the same time that Darwinism ran aground as described in chapter 7.

Starting with Copernicus, after almost 500 years of extraordinary progress, advances in fundamental physics ground to a halt because in the early decades of the twentieth century, physics ran off the rails.

᷍ ᷍ ᷍

One of the major intellectual products of the Enlightenment was a philosophy called positivism (from posit – meaning to postulate, speculate or hypothesize, as opposed to *positive*—as in mathematics), a materialistic system of thought popularized in Europe during the first half of the nineteenth century by French philosopher August Comte. Comte's ideas initially were scientific in nature but would come to have an important impact on political and sociological thought (more on that in chapter 13). In the twentieth century, his ideas would also provide the basis for revolutionary new ways of thinking about physics.

Probably the leading scientific disciple of positivism during late nineteenth and early twentieth centuries was the Austrian physicist Ernst Mach (1838-1916), namesake of the *Mach number*, the ratio between the speed of the aircraft and the speed of sound in air, used as a convenient measure of speed for supersonic aircraft, as in "Mach 2" (been there, done that). Mach advocated for a radically humanistic view of science that connected ideas about *existence* to human observability and sensory experience. Within this view, only those things that can be observed and which produce empirically verifiable results can qualify as being *meaningful* where the meaningfulness of the thing itself is established by its operational utility in performing calculations that correspond to observable and measureable parameters. Can it be used in making useful calculations? If so, it is meaningful. If not, by definition it has no meaning. Within the positivist view, reality was a concept without meaning or relevance, a rather different view from that of Galileo, Kepler, Bacon, Descartes, Newton, and so many others who were all *realists.*

Positivism reached the peak of its formal development in the 1920s through the efforts of a group of scientists and philosophers known as the *Vienna circle* who, as you might suspect, met in Vienna. Due to the emphasis by this group on formal logic and mathematics, the name *logical positivism* was adopted.

In the 1920s, this group had a strong influence on quantum mechanics during its critical formative years. It seems that Einstein at least entertained the positivist way of thinking, but he would eventually reject it because in his heart of hearts, he was a *realist*. The same was true for Schrodinger and others. Niels Bohr, an atheist, was the principal advocate for positivism.

The ultimate objective of the Vienna circle in its formulation of logical positivism was to develop a theory of knowledge based on radical empiricism which eliminated all traces of ethics, moral philosophy, metaphysics, and religion from human reasoning. At its root, it was a deductive system with clearly stated materialistic presuppositions. The central concern of the group was nothing less than to establish a revolutionary scientific foundation for *all* knowledge, thus fulfilling Comte's vision of a "religion of humanity." Trace Comte's "religion" to the present, and you get humanism. Trace it back to its roots, and it takes you back to antiquity, to Greece and Epicurus. In its conception, its intent, and its implementation, positivism was and is an atheistic system.

Realists (who gave us classical physics) believed that the natural order we perceive is objectively real and that it exists independent of both the opinions and the influences of man. For realists, the objective of science is to discover physical reality as it is, as it was created by God. Today, through the methods of the natural sciences, we have gained a reliable understanding of that natural order in terms of the fundamental laws, principles, and mechanisms that account for its behavior, all of which, in the mind of the realist, are *real parts* of a *real* world. For the realist, it is taken as a given that there is a direct correspondence between that which can be empirically observed and that which actually exists. This teleological view of the realist allows him to conclude that reality can be known where that knowledge adds a level of inspiration and energy to his efforts in seeking an understanding the natural order that the positivist can never experience. For the realist, the electron is *real,* and mathematical physics and the inductive methodological tradition of classical physics are the only valid concepts for scientific methodology.

Positivists (who gave us quantum mechanics) reject the notion of an objective reality and replace that notion with a subjective empirical-mathematical system that considers only those things that can be logically or empirically verified as meaningful or suitable for use in describing some feature of the material world. The point is not to describe anything that is "real" but rather to provide consistent calculations that relate various observable parameters that can be quantified through empirical measurements. For instance, for the positivist, there is no such entity as the "electron" where such terminology is useful only as an abstraction that provides a convenient way of talking about calculations. Mathematical physics and the inductive methodological tradition of classical physics have no place within this radically empirical and humanistic system. I say "humanistic" because of the interdependent organic connections within the *Machian* system between scientific observation, human sensory experience, and the natural world. If that sounds a little weird, then you are on track

∽ ∽ ∽

Please bear with me for the discussion that follows, which entails a somewhat simplified review of some basic aspects of atomic physics and cosmology. The purpose is to show that the problems in atomic physics are fundamental, and for physicists, they are as troubling as they are fundamental—or at least they should be. There is very little hope that these problems will be resolved within the current paradigm. We will address atomic physics first and then cosmology.

Quantum theory treats the free electron (an electron in free space and not bound in atomic structure) as an abstract mathematical point to which is assigned physical properties of mass, angular momentum, and electric charge where the properties of these parameters are determined empirically. The proton and the neutron are treated in similar fashion. Given the very small size of the subatomic particles, this approach to the modeling of the atom seems perfectly reasonable and fits nicely into the simplistic, easily visualized Bohr planetary model of the atom with electrons in orbit around the nucleus. The simplest atom, the hydrogen atom, could then be viewed as a single electron in orbit around a single proton, except that under quantum theory, the position of the electron in its orbit is not precisely known, *cannot* be precisely known, and is represented as a probability "cloud," a distribution of probable positions at any given time.

205

It would be expected that a comprehensive theory of atomic physics would inherently account for the fundamental properties of atomic structure such as the physical stability of the hydrogen atom and why its bound electron under acceleration does not radiate, which if it did would result in loss of energy and decay of the electron into the nucleus of the atom. How do we account for quantization of the excited state energy levels of the hydrogen atom, the magnitude of which are describable in such simple mathematical terms as the Balmer and Rydberg equations? Physicists can provide no *physical* explanation that accounts for such fundamental properties of atomic structure, even for the simplest element, the hydrogen atom. These three big questions concerning *stability, radiation, and quantization* remain unexplained in terms of physical laws and principles within the present paradigm of quantum theory. But there are two additional questions that could also be added to this list.

One such question pertains to the so-called *ground state* of the hydrogen atom, the state below which the electron "cannot" go. If the present ground state—or more appropriately the *minimum energy state*—is in fact a physical limitation of the hydrogen atom in the same way that the speed of light is a limitation of wave mechanics, then there must be a *physical* explanation for why that limitation exists.

The other question pertains to the inability of current particle theory to account for the *spin properties* of the free electron, for example. If a particle has angular momentum, a mechanical property, then something mechanical has to be "spinning," and it is impossible for an abstract mathematical point, which has no extent in space to have angular momentum, no matter how much mass you want to assign to that particle or how fast it is "spinning." That is why the angular momentum of the electron is described as *intrinsic*.

So, atomic physics under the quantum paradigm is unable to give an account of the fundamental properties of the hydrogen atom concerning *stability, radiation, quantization, minimum energy state,* and *electron angular momentum*. A comprehensive understanding of atomic physics *must* give a consistent account for all *five*.

But this is just the beginning. Physicists also discovered that the quantum world exhibits bizarre features and presents mind-bending paradoxes. What are we to make of a theory that tells us that particles exist everywhere at once, or that *effect* can precede *cause*, or that the physical state of an electron is indeterminate until observed through some measurement made by a human external agent, hence, the "organic" connection? And what is going on with physics when the classical notions of *particle, position, and motion* are supplanted with the abstract nonphysical mathematical notion of a *probability cloud?* And the *big* question: just what is it that accounts for the boundary between the quantum world and the world of classical physics? Can we "stand" (so to speak) at that boundary with one foot in the quantum world and the other foot in the classical world (so to speak)? And why do we see this fundamental division within the material world in the first place? Given such reasoning, we should not be surprised that things got a little weird.

The *weirdness* we perceive does not reside within the natural order itself. It resides in our minds as representations of that order as provided through what came to be known as the *Copenhagen interpretation* of quantum mechanics. And it was logical positivism with its ability to accommodate error, ignorance, chance, and irrationality that provided an epistemological home for these ideas. Physics did in fact run off the rails, and it was logical positivism that accounts for that derailment.

∾ ∾ ∾

We now turn to cosmology wherein astronomers and astrophysicists do their work principally under the paradigm of classical physics and Einstein's general relativity. Einstein's general theory points to a Big Bang creation event some 13.8 billion years ago. While Big Bang cosmology is broadly (if not warmly) accepted, there are, nevertheless, a number of quite serious problems. I address just three of these here, but if these three alone could be resolved in a consistent manner, it would constitute a significant advancement in the cosmological sciences and quite possibility a radical revision to our understanding of cosmic reality.

(1) Within Big Bang cosmology, our best estimates of the age of the universe and its present expansion rate suggest that the universe had to have expanded very rapidly (at a rate far greater than the speed of light) at the beginning to a large percentage of its present size within a tiny fraction of a second, something like 10^{-32} sec, a period called *inflation*. Further, it is believed that the ordering of cosmological structure and the low entropy state (high state of order) of the universe were established during this tiny, tiny, tiny period of time. But all of this is largely speculation necessitated by the assumption that the Big Bang initial conditions were the impossible state of zero volume, infinite density, and infinite energy density.

(2) We have discovered dark matter and dark energy, which make up around 95 percent of the mass of the universe, but astronomers and astrophysicists do not know what they are.

(3) There seems to be some mysterious source of energy that drives the Sun's dramatic temperature inversion going from 5800 K at the photosphere (visible surface) of the Sun to almost 2,000,000 K in the corona, which, assuming that the sole source of the Sun's energy is the nuclear reactions at its core, is contrary to the most fundamental principles of thermodynamics.

The fundamental nature of these problems suggest that there is little hope that they can all be resolved within the context of current Big Bang cosmology and quantum mechanics.

When viewed from the perspective of classical physics and the world as we normally perceive it, the quantum world is simply bizarre. Nevertheless, let us give credit where credit is due. We cannot discount the extraordinary intellectual achievement of quantum physicists in providing a comprehensive description of the quantum world, no matter its paradoxes and weirdness or its limitations. We must acknowledge this contribution to modern civilization in that most of the technological advances of the twentieth century rest on quantum mechanics. In the end, however, quantum mechanics can be viewed as an elaborate *mathematical device,* albeit a very useful and powerful device, but by its own positivist epistemology, it cannot be real. And therein lies the problem.

By late 1920s, it became apparent that atomic physics was being faced with one of two options. From the perspective of the realist, physicists should question the basic assumptions of quantum theory and pursue an understanding of atomic physics within the framework of classical physics and its inductive methodological tradition. From the perspective of the positivists, it was just a matter of accepting the enigmas and paradoxes of the quantum world and interpreting findings in terms of positivistic conventions and propositions, and it was this school of thinking (the *Copenhagen interpretation*) that, under the considerable influence of Niels Bohr (who was from Copenhagen), won out and eventually became the consensus view of quantum mechanics.

However, at the same time that quantum mechanics was being accepted and formalized, logical positivism was encountering problems. Ultimately, the glorious age of positivism was short-lived where the first signs of crisis were already emerging in the early 1930s. By 1960 or 1970, logical positivism as a legitimate philosophy had officially come to an inglorious and embarrassing end when philosophers of science Karl Popper[3] and Thomas Kuhn[4,5] pointed out that logical positivism, in its desire to rid the world of the metaphysical, was itself a metaphysical system. With poetic justice, it was condemned by its own theory of meaning. It was self-refuting.[6] Philosophers really hate it when that happens.

Thus, logical positivism as a philosophical system died of natural causes, imploding on itself in a fitting end—a little *phtttt* and it was all over. Nevertheless, for good or for evil, its seeds had taken root in the minds of scientists across the spectrum of the physical and biological sciences. There it lives on as *scientific materialism.*

৵৹ ৵৹ ৵৹

To gain some sense of the radical impact positivist philosophy has had on common-sense reasoning and the natural sciences, we refer to the concerns of particle physicists Lee Smolin who is deeply troubled that there have been no substantive advancements in physics in the last forty or fifty years, and he wonders why. Apparently, it never occurs to him that it may be due to the burden of the quantum paradigm that rules scientific research today. In the introduction to his book, The *Trouble with Physics* (2007),[7] he laments over the futility of *String Theory*:

> On the other hand, if string theorists are wrong . . . then we will count [them] . . . among science's greatest failures, like those who continued to work on Ptolemaic epicycles while Kepler and Galileo forged ahead. Theirs will be a cautionary tale of how not to do science, how not to let theoretical conjecture get so far beyond the limits of what can rationally be argued that one starts engaging in fantasy.

String theory provides a striking example of the influence that positivism has had on physics where, as Smolin suggests, string theorists may have "let theoretical conjecture get so far beyond the limits of what can rationally be argued that one starts engaging in fantasy." There is not a *thread* of physics in string theory (pun intended). Smolin's point in his book is that for roughly fifty years now, there have been no significant advances in physics. None of the basic problems in physics that existed fifty years ago have been resolved, and there is little hope for progress in the future.

Smolin's criticism of his profession is severe, and his analogy concerning competing second- and sixteenth-century theories of astronomy is appropriate. It also illuminates the dangers of uprooting scientific methodology from the traditional inductive methods of Bacon.

Within positivism, there is no place for such a concept as transcendent truth. The Vienna circle of positivists would say that there is no verifiably meaningful way to prove things such as statements like "God loves us." Such a statement is just as nonsensical to the positivists as the nonsensical statement that "fear is the square root of indigo" is to the realist.[8] Such is the nature of the great divide between these two views of the world. They speak the same "language" (sort of), they look at the same world and the same evidence, but they discern meaning through very different "eyes."

Positivism failed as a rational theory of knowledge. Nevertheless, its irrational conceptions, once embedded in quantum mechanics, took on a life of their own. To physicists, it no longer mattered that positivism as a philosophical system had self-destructed. After decades of incubation in secular universities, it became dogma.

Thus, blinded by its positivist assumptions, physics ran off the rails, and rather than respond to the ensuing crisis by reexamining its fundamental assumptions and returning to the unfinished work concerning *stability, radiation, quantization, minimum energy state*, and *electron angular momentum*, physicists stubbornly defend their commitment to a positivistic theory and all its *unreal* baggage in a very *real* world.

∝ ∝ ∝

Realism gave us classical physics. Positivism gave us the quantum mechanical theory of atomic physics with its burden of enigmas and paradoxes and its inherent inability to give a *physical* accounting of fundamentals, such as atomic and molecular structure. Einstein rejected the Copenhagen interpretation. His dream for a unified field theory envisioned[9] a "programme which may suitably be called Maxwell's . . . His vision called for "start[ing] with a classical field theory, a unified field theory, and demand[ing] of the theory that the quantum rules should emerge as constraints imposed by that theory itself." Einstein spent much of his later years in search of such a theory but failed. He went to his grave, however, under the conviction that "God does not play dice."[10]

As for the quantum world, it appears that a bit of "reality" is finally beginning to creep in. For instance, discontent at the highest echelons of physics was publicly expressed recently by none other than Nobel Laureate Steven Weinberg who used to be happy with quantum mechanics but now is not so sure. In an online article in *ScienceNews*,[11] editor Tom Siegfried reports on Weinberg's remarks at a meeting in October 2016 in San Antonio. Weinberg discusses the growing concerns about quantum mechanics among quantum physicists and speculates about a "deeper" quantum theory or maybe a replacement theory—or maybe not. Or maybe there is a surprise in store. Weinberg says:

> There's always a third possibility that there's something else entirely, that we're going to have a revolution in science which is as much of a break with the past as quantum mechanics is a break from classical physics. It may be that a paper from a

graduate student tomorrow morning will lay it out. By definition I don't know what that would be.

But there is a fourth possibility that Weinberg failed to consider, probably because it is unthinkable. Maybe classical physics was the correct
theory after all. Maybe the revolution Weinberg speaks of will consist
simply of returning, as painful and humiliating as it might be, to classical
physics and a realist theory of knowledge. Maybe then we will find out
what went wrong. Who knows, maybe quantum theory was simply constructed on a foundation of bad assumptions. It is not at all difficult to see
how that might have happened under the positivistic theory of knowledge.
Paradoxically, it seems that positivists, who abhor realism, are dissatisfied
with quantum theory because it is not "real" enough. Regardless, we can
be sure that physics will eventually find its way "home" to reality in spite
of the present dilemma.

In the end, it may turn out that within the long saga of the natural sciences, quantum mechanics might simply be viewed as classical physics'
version of the *Prodigal Son*. Rebelling against its roots, it had its fling
in a far country. Having worn itself out and exhausted its resources, it
is time to come home to its roots in classical physics, Bacon's inductive
methodological tradition and *realism*. On that day, Einstein's dream will
be fulfilled, the quantum paradoxes will be resolved, physics will be unified, and physicist Lee Smolin will be all grins. Steven Weinberg? Not so
sure, but he will let us know. And on that day we can begin to explore the
universe in search of that which for the last 100 years was sitting there
in plain sight, hidden beneath the confusion of quantum physics and the
positivist paradigm.

You don't have to be a prophet to know that these things are true, but you do have to be a *realist* even if that means that in doing so you are giving an intellectual assent to the idea of a Creator. Just suck it up. If the early Christian natural philosophers had embraced the positivist view of the world, modern science would never have, and never could have, developed. Realism will prevail. Just wait and see.

❧   ❧   ❧

In early twentieth century, many physicists—but not all—were persuaded to abandon realism (and in some cases, their Christian beliefs) and embrace a radically different epistemology. They became positivists, and under the influence of positivism, atomic physics went off the rails. The result was the Copenhagen interpretation of quantum mechanics with its weirdness, paradoxes, and uncertainties. As for those physicists who still held to their Christian faith (and there were many), while by virtue of their faith they did not abandoned realism, they did, however, accept the Copenhagen interpretation as it was delivered to them through their formal education. Some sought to find a way to "fit" Copenhagen into their Christian beliefs through narratives that sound somewhat like (no offense) Darwinian "just-so" stories. Most probably did not bother.

Both positivists (atheists and agnostics) and realists (theists and deists) who accept the Copenhagen doctrine attempt to find a rational place for chance, uncertainty, and even irrationality[12] in their view of the world. The positivists try to fit *chance* into an otherwise highly *ordered* world in their attempt to explain the world by *chance* and *law* alone, even to the point that they expect that ultimately classical physics and quantum physics

215

can and will be unified in a coherent and complete naturalistic scientific account of the world.

The realist tries to fit *chance* as a natural cause into a teleological account of the world. Some suggest that the quantum world was necessary because that was one of the many "hidden" ways in which God influenced the world. Such an appeal fits nicely into the popular eighteenth-century English poet William Cowper's poem[13] "God moves in a mysterious way, His wonders to perform." But they would all admit that these attempts at reconciling quantum physics and Christian faith are only theological speculation and nothing more.

Both the realist and positivist epistemologies entail *mystery*, which is an attribute of almost all religions. Thus because of its aura of mystery, you could say that quantum mechanics has a certain subliminal religious appeal. It is certainly true that leading quantum mystics love their role as high priests in extoling the "mysteries" of the quantum world to us common folk.

∾ ∾ ∾

In spite of the technological achievements of the twentieth century developed under the quantum paradigm, advances in physics itself ground to a halt, and it did so because the materialistic presuppositions of the positivist theory of knowledge do not "fit" the natural world. If your theory of knowledge does not "map" into or describe the real world in some fundamental way, then you should not be surprised to see strange things showing up in the theories that result. When your reasoning leads to screwy results, it may be because there is something screwy about your reasoning, not reality. Imagine that!

From this perspective, we can now make some sense of the present circumstances. The reason that we have two distinct domains of physics—classical and quantum—is because we have two distinct theories of knowledge—realism and positivism—based on two distinct views of the world—teleological and material. The teleological system gave us classical physics. The materialistic system gave us Darwinism and quantum mechanics. The ideas that the teleological and materialistic views, or the realist and positivist epistemologies can or should be unified are absurd. Likewise, the idea that classical physics and quantum mechanics can be unified, is absurd. There *is* no partition between the quantum world and the classical world because there *is* no quantum world. There is only one world, and it is 100 percent real and classical, through and through.

The future of science rests with classical physics and the teleological view of the world. The sooner we understand that, the sooner we will get physics (and a lot of other stuff) back on the rails, and when that happens, hang on! Things are going to get very interesting.[14]

CHAPTER 13

THE AMERICAN REPUBLIC

T he founding of the American Republic in late eighteenth century was an event whose time had come. This was perhaps the first time in history that a people had an opportunity to build a nation from scratch on a vast, largely unpopulated continent with undreamed-of natural resources. It is no coincidence that the development of this new nation proceeded in parallel with the development of the natural sciences, and neither is it coincidental that both were rooted in the Christian view of the world and its central doctrine of a Creator. For most of the first two hundred years of the life of the American Republic and perhaps for the first time in history, there was an underlying harmony, a *connectedness* between the religious faith of the people, the moral philosophy of the people, the political philosophy of government, and the natural sciences. This truly was a remarkable period in a remarkable place in the history of man. In the twentieth century, however, things began to change.

Major upheavals occurred in both American culture and American government during the last half of the twentieth century that had their roots in materialistic ideals that emerged in Europe during the eighteenth and nineteenth centuries. The teleological presuppositions of religious faith, moral philosophy, political philosophy, and the natural sciences in America were being threatened by naturalistic ideologies that had been advanced in Europe primarily by nineteenth-century philosophers Georg Hegel in Germany and August Comte in France, preparing the way for Darwin and his naturalistic account of the origins of man. Just as the Christian/teleological view shaped political philosophy in America in the eighteenth and nineteenth centuries, so the Darwinian/materialistic view influenced political philosophy in Europe and America in the twentieth century and continues to do so today. If we consider the progress of government, constitutional law, and moral philosophy within the American Republic during the last seventy years, we plainly see the effects of this transformation. While there is an unmistakable drifting away from Christian faith and moral philosophy, it is equally unmistakable that there remains an unshakable foundation of faith among the people that is here to stay. As to which of these two views will ultimately prevail in the hearts of the people remains to be seen.

The result today is a nation with deep political and cultural divisions. But while the political struggle goes on, the ultimate outcome of the underlying metaphysical battle is predetermined, not by superior political strategy of one side over the other, but by the facts of nature. As Michael Denton said in the closing remarks of his book *Evolution: a theory in crisis*[1]:

> The truth is that despite the prestige of evolutionary theory and the tremendous intellectual effort directed towards

reducing living systems to the confines of Darwinian thought, nature refuses to be imprisoned.

You might say that the *music of the spheres* was composed at creation, and nature refuses to dance to any other music. She *knows* her Master.

It is worthwhile to revisit the founding era[2] to remind ourselves of where we came from and to rediscover our roots. In so doing, we reflect on the wisdom of those who set down those precepts, establish a point of reference by which we may judge our own times, and gain perspective in trying to understand the nature of the transformations that are taking place today in the American Republic. Who knows? Maybe with some perspective rooted in reality, we might decide to do something about it.

ട ട ട

The opening paragraphs of the Declaration of Independence read:

> When in the Course of human events, it becomes necessary for one people to dissolve the political bands which have connected them with another, and to assume among the powers of the earth, the separate and equal station to which the Laws of Nature and of Nature's God entitle them, a decent respect to the opinions of mankind requires that they should declare the causes which impel them to the separation.

> We hold these truths to be self-evident, that all men are created equal, that they are endowed by their Creator with

certain unalienable Rights, that among these are Life, Liberty and the pursuit of Happiness.—That to secure these rights, Governments are instituted among Men, deriving their just powers from the consent of the governed.

America was founded on the natural law principles written into the Declaration of Independence. Natural law holds that there is an eternal transcendent God who is Creator and Ruler of the universe. He governs the universe through an eternal, immutable law that distinguishes between what is *just* by nature or merely "just" by the opinions of men. Man's positive laws may change from time to time and from place to place, but natural law is universal, eternal, unchanging, and true. *Only those positive laws of man that are rooted in God's natural laws are valid laws for humanity.* This was the classical understanding of natural law taught by St. Augustine and St. Thomas Aquinas, and by Dr. Martin Luther King Jr.[3] as well.

It was under the authority of natural law that the Founders declared their break with Great Britain and assumed "among the powers of the earth, the separate and equal station to which the laws of nature and of nature's God entitle them." Under natural law "all men are created equal and endowed by their Creator with certain unalienable rights," and these natural rights are secured by instituting governments among men that derive their just powers from the consent of the governed. The Declaration of Independence was written by the Founders specifically to "declare the causes which impel them to the separation" from Great Britain, but its principles are universal. It was more than just a declaration of America's break with Great Britain. It was a *universal* declaration, based on natural

221

philosophy, of the liberties of mankind. It was, and is, a *natural* religion *revealed by nature* to all mankind.

The Declaration set down four fundamental precepts of natural law: the supremacy of the Laws of Nature and of Nature's God, the equality of all men in the eyes of God, the natural Rights given by the Creator to all men and the place of government ordained by God in protecting the natural Rights of men.[4] These precepts also define the proper relationships that should exist between the people, their God and the state.

> *Under natural law, the people serve God, and the state serves the people. The people honor the God they serve through obedience and worship. The state honors the people it serves by honoring the God they serve and guarding the natural rights given by God to man.*

These are the *first things* of the American Republic, and the highest obligation of government is to honor and protect those *first things*. Under natural law and in respect of man's natural rights, no man or body of men, no earthly law or court, and no government and no positive law of man can stand in judgment of these *first things*, which are universal, eternal, unchanging, and true. These are the bedrock principles upon which the American Republic was founded. Everything about our Republic—its Constitution, its government, and its laws—are irrevocably connected to these fundamental ideals.

✧   ✧   ✧

The purpose of this brief exposition on the American Republic is to provide a backdrop against which we can evaluate the ideas of concern to this treatise. If Darwin got it right and the natural sciences in fact lead us to a naturalistic understanding of the world, then reason demands that we accept that the American Republic was founded on false premises, and it makes sense that we should seek to reestablish our nation on some other foundation rooted in reality rather than mythology, perhaps one based on whatever principles and truths might be determined through the natural sciences. There are forces at work in our culture to do just that.

But if in *reality*, the natural sciences confirm the teleological view of nature and are found to be in harmony with the revelations of the Scriptures, then Jefferson's appeal in the Declaration of Independence to "the Laws of Nature and of Nature's God" and his declarations concerning the natural law principles of liberty and the natural rights of man can be the only reliable foundation for a free people. And if the natural sciences do confirm the views of the Founders, then as a nation we must acknowledge these fundamental truths, reaffirm our commitment to our founding principles, begin the process of rooting out the present evil, and seek to heal our land.

❧ ❧ ❧

The Founders of the American Republic sought to form a new nation in which the common man and his family might enjoy civil and religious liberties, prosperity, and peace. The challenge before the Founders was how to establish a nation with a government built on these *first things*. Fortunately, they had help from political and legal philosophers such as

William Blackstone, Montesquieu, and John Locke of the seventeenth and eighteenth centuries, the example of Anglo-Saxon laws of the Middle Ages, the writings of Cicero, Polybius, and Cato of the Hellenistic period, and the Old Testament system of laws that governed the lives of the ancient Israelites.

They did their homework. They found that the history of civilization has largely been an endless struggle between anarchy and tyranny. Pure democracies have short lives, and violent deaths always collapsing in anarchy. While tyranny may be despised, it is preferred over anarchy. At least under tyranny, there is order and some semblance of safety. Then there has always been the hope that someday a powerful, wise, and benevolent savior-king might rise up against tyranny and with a mighty sword establish justice in the land, bringing peace, equality, and prosperity to the people. As for religion, they found that governments based on church-state alliances produce civil and religious *masters,* not civil and religious *liberties.* In the church-state alliance, government is bad for religion, and religion is bad for government. Like a bad marriage, they bring out the worst in each other, and that always spells bad news for the people.

These things were fundamental to the thinking of the Founders when they established the American Republic and any interpretation or description of our nation that does not have these central facts as its foundation constitutes a grave misrepresentation of just what it was that the Founders did.

❦ ❦ ❦

Much of the discussions of the Founders focused on this question: "In light of that most fundamental of human relationships, that which exists between a man and his God, what form of government would provide the best opportunity for a man to fulfill the purposes for which he was created?" They concluded that these purposes would best be served by a republican form of government.

History taught that there were three basic principles that were essential to establishing a nation in which civil and religious liberty could prevail. The first was that because of the corrupting influence of *power*, a republican form of government was required with a built-in system of checks and balances between the Executive, Legislative and Judicial branches of government. The second was that government should be given only those powers necessary to its civil duties and none other. The third was that the form of liberty the founders envisioned for this republic would be impossible without virtue among both the people and their leaders, and that virtue could only be sustained through religion and not just any religion. When they spoke of "religion," the religion they had in mind was that of Protestant Christianity.

Professor Daniel N. Robinson of Oxford and Columbia Universities commented on the importance of virtue in the American Republic in the first lecture of his Great Course lecture series[5] "American Ideals: Founding a 'Republic of Virtue.'"

> One of the works that was widely read by them [the Founders] was Montesquieu's *Spirit of the Law* in which Montesquieu talks about the different kind of personality you have to develop depending on the sort of regime under which people will live. In a tyranny, he says, "what you have to

cultivate in the people is the capacity for fear. In a monarch, you have to cultivate a taste and capacity for honor. But in a republic," he says, "what has to be cultivated more than anything else is virtue, public virtue."

The Founders had much to say about the importance of virtue. Here is a brief compilation from W. Cleon Skousen in his book *The Five Thousand Year Leap*[6]:

> Only a virtuous people are capable of freedom. As nations become corrupt and vicious, they have more need of masters.

> I think with you, that nothing is of more importance for the pubic weal than to form and train up youth in wisdom and virtue. Wise and good men are in my opinion, the strength of the state; more so than riches or arms.
>
> Benjamin Franklin

> Of all the dispositions and habits which lead to political prosperity, religion and morality are indispensable supports. . . . reason and experience both forbid us to expect that national morality can prevail in exclusion of religious principle.
>
> George Washington

> But neither the wisest constitution nor the wisest laws will secure the liberty and happiness of a people whose manners are universally corrupt. He therefore is the truest friend to the

liberty of this country who tries most to promote its virtue, and who, so far as his power and influence extend, will not suffer a man to be chosen into any office of power and trust who is not a wise and virtuous man.

Samuel Adams

Our Constitution was made only for a moral and religious people. It is wholly inadequate to the government of any other.

John Adams

❧ ❧ ❧

In the decades leading up to the Revolutionary War, the Great Awakening,[7,8] inspired through the preaching of Jonathan Edwards and George Whitefield, swept across the colonies. Its emphasis on repentance and conversion to Christ as a singular event in the life of the believer, and the two great doctrines of Protestantism, *justification by faith and the priesthood of believers*, ignited the fires of liberty in the hearts of the colonists. This spiritual *awakening* led the colonists to the realization that independence entailed more than just breaking *political ties* to the Crown. It also entailed breaking *religious ties* with the Church of England. That was the game-changer. It was this melding of the *religious* and the *political* in the minds of the colonists under the influence of evangelical Christianity and Enlightenment reasoning that inspired the American Revolution. And it was in the pulpits of the Presbyterian and Congregationalist churches of New England that the fire was started spreading this gospel of liberty like wildfire throughout the colonies.

Not only did the Great Awakening prepare the colonists for war, when it came time to go to war, it was the clergy of those "awakened" churches that left their pulpits to join the ranks of George Washington's Continental Army, taking with them patriots in great numbers, many from their own congregations. Because of the black robes many of these clergymen wore in their pulpits, they were contemptuously labeled by British loyalists as "the Black Brigade." During the Revolutionary War, many of the "Black Brigade" were hunted down and killed and their churches burned to the ground by the British, such was their contempt for these great American patriots. They were not just pastors. They were warriors, pastor-warriors.

Thus this great Christian awakening provided not only the moral justification and the rationale for the Revolutionary War, it also provided the passion and much of the manpower. Without the Great Awakening, there simply would have been no Revolutionary War and no United States of America.

◈   ◈   ◈

In 1783, after over eight long years of war, George Washington's Continental Army (with considerable help from the French) defeated the British. During the war years, the nation had operated under the Articles of Confederation which had served its war-time purposes but was completely inadequate for the functioning of the new nation. The so-called "United States" were not united at all and were in serious danger of imploding. A new constitution was required. The two greatest challenges of the Constitutional Convention that assembled in Philadelphia in 1787 were to decide the division of power between the federal government and the states, and to determine what the relationship should be between government and religion. Religious liberty was paramount.

Many viewed the victory over the British on the battlefield as a miracle. After the Constitutional Convention began its work, it became apparent that it would take perhaps an even greater miracle to bring the states together in forming a nation in which the principles of liberty for which they had gone to war could not just be realized but preserved. We can thank James Madison for his clear vision of the necessity for a strong national government to keep the new nation from descending into anarchy.

The question as to the distribution of power between the federal government and the states was resolved in the Constitutional Convention and reflected in the structure of the Constitution itself. While the Declaration of Independence provided a blueprint concerning the relationship between government and religion, just exactly how that would be framed in the Constitution was more difficult and would have to be dealt with through an Amendment to the Constitution if such was needed.

Views varied widely on the matter of religion. First, we cannot overestimate the view of the Founder's regarding the importance of virtue. The future of the new nation depended on it. They fully understood Dr. Benjamin Rush's axiom[9] that "without virtue there can be no liberty." In Virginia, Patrick Henry thought it was of such importance that government should pay the salaries of Christian pastors and teachers and presented a convincing argument for such. John Adams and George Washington did not go that far but felt that some sort of support should be provided to religion by the government. Then there were the evangelicals such as the Baptists who had suffered persecution at the hands of established state religions in Connecticut, Massachusetts, and Virginia. The Baptist argued fervently for a complete separation between church and state. Thomas Jefferson and James Madison sided with the Baptists.

෯ ෯ ෯

On Jan 1, 1802, Thomas Jefferson, inaugurated the year before as third president of the United States, wrote a letter to the Danbury Baptists Association of Danbury, Connecticut in response to their letter in which they inquired as to his view of religious liberty under the First Amendment. In his letter, he used the metaphor of "a wall of separation between Church and State"—language not found in the First Amendment itself—to describe his view of the *effect* of the First Amendment religion clause. This terminology, introduced in First Amendment jurisprudence in 1947, did not originate with Jefferson. Rather, it was rooted in Baptist tradition with origins in 1644 with Puritan dissident Roger Williams and introduced in the Revolutionary era in 1773 by Isaac Backus, a Baptist pastor. While one would suppose that evangelical Christians (like the Baptists) and an Enlightenment rationalist (like Jefferson) would have little in common, they were in fact united in their conviction that church and state must be kept separated. This common ground made them political allies, but the consequences of this *alliance of principle* ran far deeper than politics. This seemingly improbable mixture of *faith* and *reason* that had inspired the revolution would also forge a vital and uniquely American form of Christianity that emerged in the eighteenth century that simply had no precedent in Europe.

෯ ෯ ෯

The Founding generation proposed something radically new in the history of mankind. Historically, it had been assumed that the natural order for divine authority was from God to the king concerning civil matters, from God to the high priest concerning divine matters, and from the king and the priest to the people concerning their individual spheres of authority. Now, in America, the old order of divine authority was being turned upside down. Religious liberty was being viewed as a fundamental human right, a *natural right* that is inalienable and subject to no earthly authority, priest, or king. Under this new order, God is both King and High Priest, and the flow of authority for both *civil* and *religious* matters is directly from God to the people. As for the details, most looked to the Holy Bible as the authoritative source of God's will.

Given this new ordering of authority, how do you frame a government that reflects that ordering? This was the challenge the Founder's faced in 1787 at the Constitutional Convention in Philadelphia. In writing a constitution to provide for civil government, an important question arose. Should there be a Bill of Rights that explicitly enumerates those natural rights of man that are protected from government infringement, or would it be best to be silent about those rights? Again, opinions varied. Significantly, Madison, like Jefferson and the Baptists, favored a strict separation between government and religion, but he opposed a Bill of Rights, arguing that "what Congress is not *empowered* to do, it is not *allowed* to do" arguing that there was no need for such a bill. Was religion more secure if there was simply no mention of it in the Constitution, or was it necessary that certain protected rights be enumerated? It was not clear at all which provided the most protection—or the most risk. Time, however, would tell and did so in 1947.

Regardless, when it came to concerns about religious liberty, the focus of the Founders was on "getting it right" with respect to law. Little did they know that regardless of how sound the laws might be, the greatest threat to religious liberty would eventually come through the manipulation of those laws for political purposes.

⋽ ⋽ ⋽

Thomas Jefferson, James Madison, and the Baptists shared an important moment in Virginia that was vital to shaping the course of history concerning religious liberty, first for Virginia and then for the Republic. In 1776, Patrick Henry introduced an assessment bill in the Virginia Constitutional Convention to provide financial support to "Teachers of the Christian Religion." Thomas Jefferson opposed this bill and in 1779 countered with his own bill "Act for Establishing Religious Liberty," but, with the exception of the Virginia Baptists under the leadership of Pastor John Leland, Jefferson's bill had no support.

In 1785, James Madison published one of the most important treatises on religious liberty in American history[10] "Memorial and Remonstrance Against Religious Assessments" aimed directly at Patrick Henry's assessment bill. This treatise was circulated widely throughout Virginia, and when Jefferson resubmitted his bill in 1786, it passed. Religious liberty had been secured in Virginia, and it was set down in writing. The work done in Virginia would profoundly influence the First Amendment.

During this period, the Baptists had emerged as an important political force in the Virginia debates where their support to Madison's treatise and

232

their ability to gather popular support were instrumental in the passage of Jefferson's bill.

In 1788, the newly minted Constitution was sent to the thirteen states for ratification but without a *Bill of Rights*. There was a reluctance on the part of some states, including Virginia, to ratify the Constitution without a commitment to a *Bill of Rights*. In Virginia, James Madison, while having doubts about the wisdom of a *Bill of Rights,* was well aware that there was a strong possibility that Virginia would not ratify the Constitution without a commitment to such a bill. Jefferson, writing from France, put it bluntly: "A bill of rights is what the people are entitled to against every government on earth, general or particular, & what no just government should refuse or rest on inference." Madison reluctantly conceded, but it was John Leland who cinched the deal.

In March of 1788, elections were being held in Virginia for delegates for ratification of the US Constitution. James Madison and Baptist Pastor John Leland were neighbors in Orange County, Virginia, and both were in the running as the Orange County delegate to the Virginia Ratifying Convention. Even though Madison was popular, greatly respected, and recognized as "father of the Constitution," Leland enjoyed a substantial lead, such was his influence in Virginia. There was a historic meeting between Leland and Madison one afternoon under an old oak tree in Orange County where a deal was struck. Leland would withdraw as a candidate if Madison would agree to support the *Bill of Rights*. Madison agreed. Leland withdrew and gave his support to Madison, Madison was elected, Virginia ratified the Constitution, and Madison went on to become the chief architect of the *Bill of Rights* passed by the First Congress in 1791.

It is worth noting that perhaps no one in the Founding generation other than Jefferson and Madison, saw as clearly as did the Baptists the necessity of a republic in which government and religion were absolutely separated one from the other. Persecution has a way of producing clear thinking and forging moral courage. Thus some years later in 1802, when Jefferson wrote to the Danbury Baptist Association in response to their inquiry concerning his views on the religion clause of the First Amendment, he astutely answered them using their own language:

> Believing with you that religion is a matter which lies solely between man and his God; that he owes account to none other for his faith or his worship; that the legislative powers of government reach actions only and not opinions, I contemplate with sovereign reverence that act of the whole American people which declared that their legislature should "make no law respecting an establishment of religion or prohibiting the free exercise thereof," thus building a wall of separation between Church and State.

As we have seen, Jefferson did not pull the phrase "a wall of separation between Church and State" out of thin air. It already had a substantial pedigree in the New World, and Jefferson was quite clear about what he meant by it. He is sympathetic to the concerns of the Baptists. Any interpretation as to what was intended by Jefferson and his metaphor of a "wall of separation" cannot be construed apart from the context of the Baptist's letter to Jefferson. By its very nature, government, when left to itself, "encroaches." Unconstrained power seeks more power. It is an addiction.

The purpose of the First Amendment was to protect the religious liberty rights of the people from government encroachment.

Of course, the idea of a "wall of separation" works the other way as well (protecting state from church), and that is a legitimate concern, but you can't get that side of it out of the language of the First Amendment itself or from Jefferson's letter, not even a hint. The colonists went to war to gain freedom of religion from an oppressive governmental system of church and state, not to protect government from religion.

Much is made of Jefferson's letter and his "wall of separation," but what was actually at the heart of the letter from the Danbury Baptists as well as Jefferson's reply was the question as to the source of the rights of the people to religious liberty. Were those rights given under the Constitution as "favors granted" or by God as "inalienable rights"? Jefferson relieved their concerns by his reference to the natural rights of man when he went on to say in the *very next sentence*:

> Adhering to this expression of the supreme will of the nation in behalf of the rights of conscience, I shall see with sincere satisfaction the progress of those sentiments which tend to restore to man all his natural rights, convinced he has no natural right in opposition to his social duties.

This was the whole point of the Danbury letter. If the courts want to look to Jefferson for a substantive extra-Constitutional opinion from the founding era on the First Amendment religion clause, they should look to his affirmation of the natural rights of man to freedom of conscience and a "wall," protecting church from encroachment by the state.

We see a common theme running throughout the American founding. The Declaration of Independence was based on natural law principles and the natural rights of man, the Constitution was carefully crafted to reflect those principles and protect those rights, and Thomas Jefferson made it clear in his letter to the Danbury Baptists that that was the specific purpose of the First Amendment. Thus Jefferson's "wall of separation" can be properly understood only in light of natural law principles and the natural rights of man. The natural rights of man are inalienable and given by God. The role of government is to protect, not infringe upon those rights—to guard the lamb, not eat it.

◈ ◈ ◈

The Founders had a deep distrust of government and knew that the only hope for both civil and religious liberty was to keep these two institutions, church and state, separated. Thus the Founders concluded that the power of government must be strictly limited to only those powers necessary to the performance of the duties of civil government, that, and nothing more. Do not give government any more power than that which it needs. This was the clear lesson of history. If we "render therefore unto Caesar the things which are Caesar's; and unto God the things that are God's"[11] then *maybe* there will be a chance for justice, liberty, peace, and prosperity for the common man.

Thus, the Founders gave us a secular[12] government established by a secular constitution, not because they wanted a government established on secular principles as an essential aspect of republican government, but as a necessary limitation on government power. However, it is one

236

thing to accept the necessary but limited secularization of government that arises when, for the sake of religious liberty, the power of the government is restricted to only those powers necessary for the functioning of civil government. It is quite a different thing to employ secularism as the underlying principle of government, which by its nature, introduces within government a general predisposition *against* faith in God, natural law, and the natural rights of man—in short, against the founding precepts of the Declaration of Independence. The actions of the judiciary after World War II provide good reason to believe that the secularism of our present times is *not* for the benefit of religious liberty but constitutes a covert *established religion* that functions within government to undermine the founding principles of the American Republic in natural law. We simply cannot find anything within the Declaration of Independence that justifies this extreme species of secularism, and if it cannot be found there, then it is alien to the principles upon which the American Republic was founded.

❧ ❧ ❧

When the Founders spoke of *religion*, they meant Christianity, and when they spoke of *church,* they meant the churches of the various Christian denominations. When they wrote that the Congress shall "make no law respecting an establishment of religion" they meant that no Christian denomination would be established as the official religion of America and as such share political power with the state. The First Amendment guaranteed that no church-state alliance, such as that which existed in Great Britain, would ever be formed in America. As for Jefferson's metaphor of a "wall of separation," it was simply his way of reassuring the Danbury

Baptists that in America, the federal government would never have any power over religion.

The purpose of the religion clause of the First Amendment was to put every aspect of religious life in America beyond the reach and control of the federal government. However, no such limitation on governmental power was extended to the states, even though James Madison felt that the greatest threat to religion would eventually come from the states rather than the federal government. The problem was that some states had already established religions of their own, and it was pretty much a given that those states would not be willing to ratify the Bill of Rights if its limitations on establishment of religion were imposed on the states.

Some states retained a state-established religion for decades following ratification of the Bill of Rights, but by 1833, all states had passed laws prohibiting religious establishment. They did, however, permit various forms of religious expression that reflected the precepts of our founding and our religious heritage.

∽ ∽ ∽

It was common practice throughout the Revolutionary War years, and even after ratification of the Bill of Rights, to publicly acknowledge the providence of God, to give thanks for His blessings, to seek divine guidance in times of national trouble, and to acknowledge our dependence on Him. Various tributes acknowledging God and our religious heritage were offered, not only on public occasions, but in monuments and inscriptions on public buildings as well. Not all of the founders agreed with such open religious expression within the federal government (Thomas Jefferson, for example), but up until mid-twentieth century, such public expression concerning God within both federal and state governments were simply part of public life in America. This religious expression was not imposed on the people by the states; rather, it reflected the almost universally accepted disposition of the people. Politically, we may have been made up of Democrats and Republicans and Independents and other political persuasions, but the great force that unified us as Americans was that we were one nation under God and not as some abstract notion, but in reality, and it was in that unity and in that reality that we maintained our strength.

Further, it is simply a fact that you cannot separate the story *of* America, the good and the bad (and there was plenty of that), from the story of Christianity *in* America. Those who settled the Massachusetts Bay Colony in early seventeenth century were not secularists; they were Puritan Christians, and they came to America seeking religious liberty, not secularism. The forces at work in the Great Awakening, which inspired the Revolutionary War, arose from the unification of the ideals of Christian faith and Enlightenment reason, not secularism. Those of the founding era, who went to war, fought for religious and civil liberties, not secularism. The opening paragraphs of the Declaration of Independence appealed to the laws of nature and nature's God and the natural rights of man as the basis of a free republic, not secularism. And it was Christians, not secularists, who persisted against the curse of slavery until it was finally ended in America. Christianity, for better or for worse, is deeply woven into the fabric of our history.

Kenneth Scott Latourette, professor and fellow of Berkeley College in Yale University summarizes the impact of Protestant Christianity under the influence of the Great Awakening in shaping the soul of the Republic.[13]

The Protestantism of the Thirteen Colonies was laying the foundations for the democracy which found expression in the American Revolution and the United States. For example, in New England the clergy were preaching the rights which come from nature and nature's God, the theory that all men are born free, the duty of resistance to encroachments on those rights, and the popular elements in government. While many of the clergy looked askance at pure democracy, the radical Protestantism which predominated in the churches in the Thirteen Colonies, seeking as it was to carry through the distinctive principles of the Reformation, salvation by the faith of the individual and the priesthood of all believers, underlay and permeated the democracy which characterized the United States.

The cultural and religious landscape in America today is radically different from that of the founding era. This is due in part to the influence of immigration in the latter decades of the nineteenth century and the early decades of the twentieth century, which introduced a variety of new religions and cultures into American life. Further, we cannot ignore the effects of the Civil War. This war was not fought on distant battlefields but on our own soil. A nation cannot endure such carnage, its own soil soaked in the blood of its countrymen, without being changed. Professor Allen C. Guetzo of Gettysburg University in his lectures on *The American Mind*[14] provides a sobering survey of this period and the devastating and lasting effect that war had on the American spirit.

But this was only part of the story. The cultural and religious changes to the soul of America that took place cannot be explained by the effects of the Civil War and immigration alone. The main cause of these dramatic changes that took place in the twentieth century was due to the influence of European Enlightenment philosophers[15] like Jean-Jacques Rousseau of the eighteenth century, nineteenth-century philosophers Georg Hegel, Auguste Comte, Friedrich Nietzsche, and Karl Marx, and the English naturalist Charles Darwin. The ideas of these and other thinkers spread during the nineteenth century throughout the universities of Europe, and in the last half of the nineteenth century and the early decades of the twentieth century, they were introduced into America's secular universities through American students seeking advanced degrees in Europe. When these new doctorates returned home to America, they brought these new ideas with them and introduced them in the curricula of America's secular universities, and it was through this means that those ideas were implanted in the hearts and minds of the educated elite of America. Many of those elite became the educational, social, judicial, and political leaders of early twentieth century. The net result was nothing less than a cultural revolution that introduced a secular worldview into the cultural and political institutions of America, which was completely foreign to anything the founding generation could have imagined. While progressives like Teddy Roosevelt and Woodrow Wilson never intended such a complete cultural and moral revolution, Thomas Dewey, Oliver Wendell Holmes, and Margret Sanger most assuredly did. The distinguishing factor between them was God. Dewey, Holmes, and Sanger were committed Darwinists as were other bad boys of the twentieth century like Sigmund Freud who, according to Benjamin Wiker,[16] corrupted our views on moral behavior and Alfred Kinsey who corrupted our views on sexual behavior.

At the heart of this transformation was the idea of "inevitable progress" conceived by German philosopher Georg Hegel (1770–1831). Hegel believed that societies and governments change in response to historical contingencies and naturally progress toward some ultimate state of perfection. He surmised that there must be "laws of progress" that govern this process and that by discovering those laws, man can use them as a guide in his pursuit of the perfected state. Apply these laws to government, social policy, and human behavior, and surely utopia will be just around the corner. However, if governments are constrained by fixed principles like those of natural law and our present Constitution, then progress will be stifled.

The French philosopher and child of the French Revolution, Auguste Comte (1798–1857), provided a historical framework for Hegel's ideas in his concept of the *three stages of the history of mankind* where mankind progresses (through an evolutionary process as Darwin would later hypothesize) from the initial stage of ignorance and superstition to metaphysical speculation in the second stage to the final stage of science and reason or, within Comte's terminology, the age of *positivism*.

Positivism, as conceived by Comte, whether applied to science, law, political systems, or moral philosophy, is a materialistic philosophy, a theory of knowledge that rejects metaphysics and theism. In Comte's world, teleology was an illusion. There was no design or intention behind nature. His system provides an appealing framework for understanding our history—where we have been, where we are, and what our future holds. According to Comte, Western civilization has been stuck for almost

2000 years in the second stage in Christianity, and the time has come for Christianity to give way to the inevitable transition to the final positivistic stage of history. According to Comte, this would be the "regime of human reason" during which he was quite sure would emerge what he called a "religion of humanity," a humanistic religion based on science and reason that would sweep over the world and of which he would be high priest. His death in 1857, however, put an end to his papal aspirations. The future of positivism would fall to others.

While Hegel was author of the concept of inevitable progress, it was Darwin who later secured its intellectual foundation (such as it was). Darwin "discovered" the "laws of progress" concerning the evolutionary development of mankind from lower forms, thus giving Hegel "scientific status," a sure indicator that mankind was progressing into Comte's final stage of history and the utopian age. Woodrow Wilson reflected on this connection between Hegel and Darwin when he expressed his view of government as a "living thing," which was to be understood according to "the theory of organic life." Constitutions are not "fixed" but should be "Darwinian in structure and in practice." Further, it was believed that mankind had progressed to the point where there was no longer need to impose such stringent limitations on government such as the balance of powers reflected in the US Constitution. It was time to discard God's timeless, fixed laws as the basis for government and embrace man's positive laws, which could be more easily changed to accommodate the times and the changing conditions of existence. The modern world that emerged out of the industrial revolution of the nineteenth century was dynamic and complex, and many believed that only a progressive philosophy of government could provide the flexibility and governing power necessary to direct and manage such a world. And probably for most leading progressives, such as Woodrow Wilson and Teddy Roosevelt, this was simply a practical consideration, which had no influence on their view of God's place in American life. For many others, however, who held to the Comteian view of progressivism, the idea of getting God out of American life was the whole idea.

A central premise of Comteian progressivism is that the rights of the individual should be determined by the state, not the Creator. The "higher good" of the state takes precedence over the natural rights of the individual, thus inverting the divine order of things. Given a nation infected with progressive thinking, all of the empirical data tell us that there will be one of two outcomes. One is suffocation by totalitarian rule. The other is suffocation by socialism. It will be "hard" tyranny or "soft" tyranny.

According to Professor Ronald J. Pestritto[17] at Hillsdale College, the American Progressive Movement—embraced by such notables as Oliver Wendell Holmes, Theodore Roosevelt, Woodrow Wilson, John Dewey, and Margaret Sanger—constituted the first attempt in America's history to overthrow (albeit through civil discourse) the natural law principles of our Founding Fathers. Given progressivism in its more Comteian form, political power inevitably migrates toward the executive who rules as a monarch over a socialist state where the rights of the people are whatever the state says they are, the people serve the state rather than the Creator, the state serves its own best interest rather than that of the people, and public virtue is measured in terms of allegiance to the state. The only place for God in this enlightened view of government is in the practical benefits of His grace and moral law that comfort people in their oppression and restrain them in their behavior, thereby providing a positive force in maintaining social order—all for the good of the state. Religion and its influence must be confined to the private sphere. Under this political philosophy, individual liberty is not viewed as an inalienable right; rather, it is subject to the needs of the state, which makes whatever provision it may judge as appropriate for the needs of the individual.

Just as evangelical Christians and Enlightenment rationalists found common ground in the establishment of the American Republic, so metaphysical materialist and political progressives find common ground in engineering the transformation of America. The new standard for personal virtue and public morality in this enlightened state will be equality, achieved through social policies directed at redistribution of wealth and a normalization of social status to the lowest common denominator as determined, of course, by the wisdom of the state. Justification for such policies is that this is what *equality* and *fairness* require. The guiding principle of social policy is *entitlement*, not *liberty*. The cost is measured in terms of liberties lost and the general degeneration of all aspects of society. Many, maybe most, might be miserable, but if that is the price to be paid to achieve equality, then so be it. There can be no utopia without equality. It just may not be the utopia you hoped for. Regardless, we should all rejoice in our common allegiance to the "greater" good as we "progress" from the old self-centered natural law concerns of *rights* and *liberty* to the new altruistic concern of social and economic "justice."

While as a practical matter, it was necessary that the Constitution be interpreted in respect of the changing conditions of the life of the nation and its people, the natural law principles that underlie that interpretation are not subject to change. Nevertheless, in the first few decades of the twentieth century, under the influence of progressive dogma, fundamental changes in the thinking of the judiciary from a system of justice based on natural law with its transcendent principles of justice were rejected in favor of a system of justice based on positive or man-made laws tailored to meet the "needs" of the situation. The effect was a radical change in the place of religion in *public life* where God might be given due respect, but his divine laws were discarded in favor of the positive laws of man.

A similar process took place in the physical sciences in early twentieth century (chapter 12) when the long-held teleological context of the laws of physics developed by Christian natural philosophers was rejected for the positivist philosophy, which holds that reliable knowledge concerning our understanding of the natural world can only be obtained through information gained through sensory experience and reason alone without appeal to teleological notions of law and principle.

Note the prominent place of the descriptor "progressive" and the reference to "the greater good" in the following statement by the American Humanist Association[18]:

> **Humanism** is a *progressive* lifestance that, without supernaturalism, affirms our ability and responsibility to lead meaningful, ethical lives capable of adding to *the greater good* of humanity. (my emphasis, italics only)

The secularization of the public sphere has been the heart and soul of progressive political philosophy since the mid-twentieth century. Today, by law, whatever government does must be secular in nature. Now that sounds okay if you accept the idea that *secular* concerns itself simply with the legitimate matters of civil government and has nothing to do with religion and certainly has no anti-Christian meaning. If this is what you think, then your thinking is politically "correct." History, however, as is often the case, tells a different story.

Secular was first coined as a word in mid-nineteenth century by George Holyoake,[19] a radical atheists and follower of August Comte. He invented this word as an aid in selling his radical atheistic ideas concerning societal revolution to a largely Christian world. Holyoake felt that *secular* is far less shocking to Christian sensibilities than *atheism*. It is a word with two faces. One cajoles while the other poisons your soul. Holyoake promoted *secularism*, but he was selling *atheism*. Thus these three concepts— progressivism, positivism, and secularism—all share roots in *atheism*. Together they carry on Comte's program preparing the world for the final stage of history and the *religion of mankind*.

❧ ❧ ❧

This revolution in culture in twentieth-century America constitutes what is, in effect, a second American war of liberation (albeit without guns). At the heart of this liberation were three fundamental ideas, which had been simmering in the secular universities of America since before the turn of the twentieth century. These ideas provide the precepts needed for the establishment of a secular society set free from our founding pre-cepts in natural law and Christian moral philosophy. The three founding principles of this liberated society are *moral relativism* where morality is reduced to personal preference, *multiculturalism* in which all cultures and beliefs are held to be morally equivalent, and *postmodernism* that rejects all notions of transcendent truth and reduces all notions of ethics and morality to social constructions.[20] These ideas are completely foreign to the natural law principles and Christian moral philosophy that have provided the foundation for civil and religious liberty in America since its founding. Today, however, they are right at home in public education, the universities, the liberal media, and a progressive judiciary.

The Founders abhorred the notion of government based on a church-state alliance and took measures to make sure that could never happen by writing protections into the First Amendment. Nevertheless, we are well on the way to transforming government into such a church-state alliance and, just like its predecessors in seventeenth-century Europe, this is an alliance that is about the consolidation and centralization of power and serves its own best interest instead of that of the people. This alliance is between secularism, the religion (which, while almost invisible, nevertheless leaves a very large "footprint" on the culture) and progressivism, the political philosophy. It *rules* the people by power it has seized, rather than *govern* the people through power it has lawfully been delegated. But if you value getting "free stuff" and being taken care of more than you value your liberties, then this is the form of government for you.

Whereas the *first* American Revolution was about freedom *of* religion for the common man, this *second* American Revolution is about freedom *from* religion—meaning freedom from God and his oppressive moral laws. Its principle political purpose is to overthrow the founding precepts of the Declaration of Independence and to invert the natural order of things where the rights of the individual are given not by his Creator but by the secular progressive state. Under this Comteian system, it will be evolutionary psychology and social Darwinism that establish the standards for right conduct and social order rather than the Creator's natural laws and Christian moral philosophy. Two foundational goals of *that* church-state system, which to some extent have already been introduced into our culture, are to rid society of the influence of Christian views concerning the sanctity of life and its strict mores on sexual behavior. While it is hard to object to the "do unto others," "love thy neighbor," "greater love hath no man," and other admirable features of the teachings of Christ, ultimately the ideals of a utopian society cannot be realized under a moral philosophy rooted in the life and teachings of Christ.

Whereas the first American Revolution gave us a government serving a God-fearing society, once all the goals of the second American Revolution are achieved, we will have a secular society, serving what is in reality that which the founders most feared: a church-state system of government. What that might look like remains to be seen, but we should hope and pray that the godless totalitarian regimes of the last century were not precursors to what is coming to America in this century.

❧   ❧   ❧

In understanding what was involved in making the transition from the founding era to our present times, we must consider not only the dramatic changes in societal views concerning ethics and morality that took place during the twentieth century but also the impact of changes in First Amendment law that arose in going from a relatively narrow view of religion within Protestant Christianity during the founding era, to essentially all of the world religions of our present times, including atheism in its various forms (naturalism, materialism, humanism, secularism, etc.) and nontheistic forms of belief (Buddhism, Hinduism, Transcendentalism, etc.) as well as the traditional theistic forms of belief. It is worth considering just exactly where *secularism* falls within this continuum.

When *religion* simply meant "having to do with the knowledge and worship of God," *neutrality with respect to religion* seemed to be an appropriate principle for secular government charged with the protection of our religious liberties. All that meant was that government, in its exercise of its delegated powers, must not favor one God-fearing religious sect over other God-fearing religious sects or for that matter, a godless sect. However, given the *inclusive* meaning of religion used at times (but not always) by the courts today, this is no longer the case. It now seems that the only practical approach to ensuring religious neutrality is simply to ban all traditional religious expression and require a purely secular government. This seems to be the path the judiciary has chosen.

This approach, however, is impossible. There can be no truly *neutral* position that government can assume because under the *inclusive* meaning of religion, secularism itself functions as a religion. If the Supreme Court can rule that prayer to Almighty God in public schools in which children simply ask for the blessings and protection of heaven on them, their teachers, and their families constitutes establishment of religion, then how can you argue that the act of abolishing such prayer does *not* constitute establishment (the power to abolish is the power to establish) of a secular religion, which by its nature is the antithesis of the *religion* you just abolished? And how can that not be unconstitutional when the First Amendment prohibits government from prohibiting free exercise of religion?

The logical structure of the religion clause requires that if there is something that you are prohibited from prohibiting, then that something is something you must leave alone. It is something that must not in any way be touched.

The judiciary created this horrendous mess when it strayed from the *first things* of our founding. It was not the "god" of secularism or the "god" of atheism that created us and gave us our inalienable rights. We must re-establish our foundation in our founding precepts concerning the *supremacy of the Laws of Nature and of Nature's God, the equality of all men in the eyes of God, the natural Rights given by the Creator to all men and the place of government ordained by God in protecting the natural Rights of men*. While these *first things* are deeply religious in nature, they cannot be reduced to "religion" and then subjected to the First Amendment and the opinions of the judiciary. *It is the other way around.* Ultimately, every provision of the Constitution, every law of the Congress, every finding or opinion of the judiciary, and every order of the Executive Branch must in some way be subservient to these natural law precepts.

Judaism stripped of *the law* is no longer Judaism. Christianity stripped of *the incarnation* and the *resurrection* is no longer Christianity. And the American Republic stripped of its founding principles in God's *Natural Law* and the *natural Rights of man* is no longer the American Republic.

෴   ෴   ෴

The fundamental precepts of the Declaration of Independence concerning the laws of nature and nature's God are fundamental to the America founding. These precepts have been rejected by many of the legal

251

philosophers and scholars of our secular universities, and consequently they have been rejected by many of the lawyers, judges, and justices who were educated in those secular universities. Under what authority and by what means did we reject our founding in natural law?

We are in the midst of a transformation, a radical overhaul of America, every bit as revolutionary as the one that established the American Republic. Positive law and a progressive political philosophy provide a different vision of America from that of natural law. We are now being transformed under this new vision into a secular progressive state in which Christian moral philosophy is rejected in favor of a humanistic philosophy rooted in Darwinian naturalism. All three branches of government are to some extent implicit in this transformation, but none has had such a radical impact as the Supreme Court, the least accountable to the people (if at all), and under the balance of powers almost invulnerable in their power. The most immediate effect has been the systematic purging of God and all things sacred from public education and the establishment one small step at a time of *secularism* as the religion that governs public life in America.

Public education, an institution of state governments, is the primary battleground for the transformation that is taking place. Centralized control of curriculum is imposed through such national movements as the Next Generation Science Standards, Common Core, and national education unions and associations. The stakes in this war are the minds of our children, and the teaching of Darwinism is the core doctrine with which our children are being indoctrinated.

ี๑   ๑   ๑

A major milestone in the "Comteian takeover" occurred in America in 1947 when a case was bought before the Supreme Court that would lead to radical changes in the way the judiciary viewed the place of religion in public life. The subject case was *Everson v. Board of Education of the Township of Ewing*.[21,22] The issue before the *Everson* Court was whether payment by public education in New Jersey for the cost of public transportation of Catholic students to and from parochial schools in which they were taught the basic principles of the Catholic faith constituted violation of the Establishment Clause of the First Amendment. The cost of public transportation of students to and from public schools was also reimbursed by public education. The Court was split (5–4), ruling that such payments were permissible, and there was no violation. Justice Hugo Black wrote the majority opinion of the Court. In support of their opinions, Justice Black and two of the four dissenting justices, Justices Jackson (of Nuremberg fame) and Rutledge, wrote lengthy historical reviews of the critical period from 1776 to 1791 in Virginia during which the Founders hammered out the relationship between government and religion as reflected in the First Amendment Religion Clause.

From a review of their opinions, it seems that the justices all understood and supported the strict separation principle advocated by Thomas Jefferson and James Madison. The justices also concerned themselves with an understanding of what the Founders intended. As to the split in the opinions of the Court, this was due primarily to two different ways of viewing Catholic schools in relation to public schools rather than some fundamental dispute over interpretation of the First Amendment. The concurring side saw Catholic schools as simply an alternative form of education from that of public schools where the cost of transportation was incidental to the cost of education itself. The dissenting side saw Catholic schools as part of a powerful worldwide authoritarian religious institution, and any money trail at all which leads from government to any aspect of that institution constitutes giving direct aid to a particular religious sect. It is very likely that Madison and Jefferson would have enthusiastically joined the four dissenting justices in their dissent.

The question before the Court had to do with deciding just where you draw the *Establishment* "line." When does the obligation of the state to provide "assistance for the general welfare" of its residents, become "government aid to religion"? It was complicated. The state of New Jersey not only provided this benefit to Catholics, it also provided it to students of other faiths who attended public schools and who also required assistance with transportation. Further, even though Catholic students attend Catholic schools, their parents still paid taxes in support of public schools. While there were pros and cons to each side, the Court ruled that the practice was permissible. While this case took the Court well into the gray area of First Amendment jurisprudence, all in all, it seemed like an acceptable decision.

But there was far more to *Everson* than a question about state reimbursement of fares for public transportation of children to private religious schools—far, far more. In the end, the question of reimbursement of cost of transportation, a local issue in New Jersey, was pretty much irrelevant to the real impact of *Everson*, which established a precedent for future courts to begin the process of a nationwide secularization of public education. Why public education? Because if you want to change the culture, this is where you start.

❧ ❧ ❧

In writing the majority opinion of the Court, Justice Black looked to Thomas Jefferson's 1802 letter to the Danbury Baptists and his metaphor of a "wall of separation" supposedly for historical insight in determining how the Court should view the relationship between church and state. But *it*—what Black concluded, not what Jefferson wrote—was a sham. In his review of Jefferson's letter, Justice Black blatantly misrepresented[23] the plain context of Jefferson's "wall" to justify *his misrepresentation* of the place of that "wall" in Establishment Clause jurisprudence. The concern of the Danbury Baptists in their letter to President Jefferson was protection of the natural rights of man from government encroachment. Jefferson's "wall of separation" was a metaphor for how the First Amendment Religion Clause provided that protection, at least from the federal government. This was the only rational interpretation that could be derived from either the Baptist's letter or Jefferson's reply. Jefferson's response was clear. It left not even a possibility of the misrepresentation contrived by Justice Black.

There was purpose to Justice Black's "madness." Justice Black's "wall," which he described as "high and impregnable" had a different purpose, a purpose that was not explicitly disclosed in *Everson* itself but which became clear in the decades to follow when several cases were decided based on the precedent established in *Everson*. We repeat from before that portion of Jefferson's reply to the Baptists concerning his "wall of separation" that makes the intent of his metaphor perfectly clear:

> Believing with you that religion is a matter which lies solely between man and his God; that he owes account to none other for his faith or his worship; *that the legislative powers of government reach actions only and not opinions*, I contemplate

255

with sovereign reverence that act of the whole American people which declared that their legislature should "make no law respecting an establishment of religion or prohibiting the free exercise thereof," thus building a wall of separation between Church and State. Adhering to this expression of the supreme will of the nation in behalf of the *rights of conscience*, I shall see with sincere satisfaction the progress of those sentiments which tend to *restore to man all his natural rights*, convinced he has no natural right in opposition to his social duties. (my emphasis)

The issue at stake for the Baptists was the sanctity of the *rights of conscience* and the *natural rights* of man. In Connecticut, the Baptists were experiencing persecution at the hands of the Congregational church (established by the Puritans) which in 1802 was the established church of Connecticut as well as of Massachusetts and New Hampshire. In Connecticut, what limited rights the Baptists did enjoy were being doled out as *favors granted* rather than the *inalienable rights* of the Declaration of Independence. Even though the First Amendment limitations applied only to the *federal* government, the Baptists wrote President Jefferson, hoping that the influence of his opinion might help relieve the oppressive conditions in Connecticut. Perhaps the "spirit" of the Religion Clauses would be adopted in good faith by the states. It turned out that eventually this is what happened but not in 1802. Regardless, the Baptists gained reassurance and moral support from the president of the United States and author of the Declaration of Independence—not too shabby.

Jefferson's principle concern was the *natural rights* of man given by God to man. Government has no authority over those rights, only a duty to protect those rights. When Jefferson responded to the Danbury Baptists concerning the natural rights of man, the context of his letter was that of our founding precepts in natural law.

> *The precepts of our founding as set down in the Declaration of Independence, are established under the providence and authority of the "Laws of Nature and of Nature's God", not the Constitution and not the courts.*

The Constitution was framed under those fundamental precepts, and no institution of government or court of men has any jurisdiction over those precepts. Justice Black took Jefferson's "separation of church and state" and recast it as a judicial principle—"separation of God and state"—to serve *other* than a constitutional principle, but in the *name* of a constitutional principle.

⤚ ⤚ ⤚

It is crucial that we understand the distinction between natural law and religion so that we are not so easily deceived when someone like Justice Black treats natural law precepts *as if* they were religion. Natural law and religion have much in common (God, for instance), but they occupy two distinct and very different places in the framework of the American Republic. God and his natural laws are revealed to us through nature and provide the fundamental precepts of our Founding as a nation, reflecting the nature of the relationships between the people, their God and their government (the *first things*). Religion, within the context of the First Amendment, concerns the worship of God by the people based on revealed knowledge. You cannot entangle one *with* the other and then treat the one *like* the other—even if you are a justice of the Supreme Court. Or, in short, the High Court cannot treat our founding precepts as if they were "religion" and then abolish them in the name of the First Amendment from every aspect of government—and in particular, public schools. No. The fundamental precepts of our founding in natural law, the *first things*, must be honored throughout government, even by the Supreme Court, even in public schools, and in spite of the fact that these *first things* come to us from the Creator.

The purpose of the First Amendment was to protect religion by placing limitations on the powers of government. When Justice Black looked to Jefferson's letter for guidance regarding *intent,* it seemed like a reasonable thing to do. But what he gave us was a deceptive version of Jefferson's metaphor of a wall of separation. In Jefferson's world, that wall protected the natural rights of man from encroachment by government. In Justice Black's world, that wall protected government from God even though there is not even a hint of a concern in the First Amendment about protecting the state from either religion or from God. It was just assumed by Justice Black.

Jefferson stood with the Baptists on one side of the wall, defending the natural rights of man. Justice Black stood on the other side of the wall, defending the government supposedly from some imagined invasion by religion. But that was just a cover. Justice Black's real target was the abolition of God from public schools. That purpose did not become clear until years later as the High Court began to systematically expel God from public schools.

❧ ❧ ❧

Everson was a landmark decision, not because of the "deep" constitutional significance of the question that had been brought before the Court, but because it was the first time a federal court had exercised its powers under the Fourteenth Amendment in applying the Establishment Clause of the First Amendment to a state, a process legal scholars call *incorporation*. When the Bill of Rights was ratified in 1791, the limitations on power imposed on the federal government to protect those rights did not apply to the states. The Founders decided (against Madison's earnest pleas) that it should be left to the states to decide how they might deal with this question.

After defeat of the Confederacy in the Civil War by the Union Army in May of 1865, the Union was faced with staggering problems in rebuilding a nation torn by civil war and there was no priority greater than emancipation of the slaves. The purpose of the Fourteenth Amendment was to make it clear that the freed slaves were to have all the natural rights (*privileges and immunities*) as native-born citizens. Here is Section 1 of the Fourteenth Amendment.

All persons born or naturalized in the United States, and subject to the jurisdiction thereof, are citizens of the United States and of the State wherein they reside. No State shall make or enforce any law which shall abridge the privileges or immunities of citizens of the United States; nor shall any State deprive any person of life, liberty, or property, without due process of law; nor deny to any person within its jurisdiction the equal protection of the laws.

259

With the ratification of the Fourteenth Amendment in 1868, these limitations were made applicable to the states. It would seem that the language of Section 1 indicates that the intent of the Fourteenth Amendment was that the act of ratification itself by the states would constitute incorporation of the individual protected rights. But that was not the case. While the *language* of the Fourteenth Amendment seems to be quite clear, it was not assumed by either the federal or the state governments that the *intent* of that language was that all the limitations on powers concerning the enumerated rights of the Bill of Rights would be instantly applied en masse to the states. You might ask, "Why not?" Good question. It might be that because of the distinctive nature of each of the enumerated rights, individual *incorporations* were required for the sake of establishing the appropriateness of their incorporation and a sound precedent written for the benefit of future courts.

Regardless, the process of incorporation did take place but did so one step at a time through individual court decisions over a period of many years in a process called *selective incorporation*. This overall process of incorporation was completed (if I am not mistaken) in *Everson*, where the enumerated rights in question were those protected by the Establishment Cause of the First Amendment.

But there was another consequence of the Fourteenth Amendment (perhaps unintended by the authors of the Fourteenth Amendment), one that led to a radical transfer of judicial power from the states to the federal judiciary effectively dividing our judicial history into two parts: *before* the Fourteenth and *after* the Fourteenth.

Prior to the Fourteenth Amendment, jurisdiction over complaints of infringement by a state of rights enumerated by that state came solely under the jurisdiction of the state in which the offense occurred and the complaint was filed. It was a state matter, and it was resolved by the state. *After* the Fourteenth Amendment and after *incorporation*, a complaint that might be filed within a particular state concerning a right enumerated under the Bill of Rights of the *federal constitution* now came under the jurisdiction of the federal judiciary and *ultimately,* under the jurisdiction of the Supreme Court, and was thus subject to appeal all the way to the Supreme Court. While this may at first glance seem reasonable and unremarkable, in reality it was a profound break from what the Founders gave us. In fact, within the range of its applicability, it essentially drove a stake through the heart of the ideals of republican government. *Now*, as a consequence of the Fourteenth Amendment and *incorporation*, whenever the Supreme Court hands down a decision concerning a case involving an enumerated right protected under the Bill of Rights, and which originated from a lawsuit filed within a state, and through appeal was brought before the Supreme Court, whatever decision the High Court might hand down, that decision immediately becomes law, not only for the state in which the lawsuit originated, but it becomes the *law of the land*, and any law in any state that is in opposition to that decision is immediately invalidated. This applies to *all* states even if the conditions and the offense that led to the original lawsuit in the originating state never existed in any of the other forty-nine states. The net result is that the other states are now subject to this new law in spite of the fact that none of those individual states had any part in enacting that law, and the interest of the people in those states was not represented because the Fourteenth Amendment/incorporation in effect stripped the people of each state of that power and transferred it the federal judiciary, and ultimately, the Supreme Court. These decisions/laws come down from "on high" rather than from legislation enacted through duly elected representatives of the people. Such law has the "look" of a tyrannical mandate, not legitimate law enacted through representative government.

The Founders would be appalled at the idea of placing such power in the hands of the federal judiciary, much less nine Supreme Court justices. Under these conditions, state and federal legislatures, state governors, and the President are cut out of the law-making process. Say goodbye to balance-of-powers and representative government. Say hello to judicial tyranny and progressive government.

In the end, the principle *constitutional* effect of the Fourteenth Amendment and incorporation was to convey power originally held exclusively by the states (and consequently, by the people) to the federal judiciary where all matters concerning the limitations of power associated with the enumerated rights of the people are now brought under scrutiny of the federal courts. This constitutes a *monumental* redistribution of judicial power from the states to the federal judiciary whereby state laws that were previously invulnerable to review on the part of the federal judiciary were now vulnerable to such review opening up opportunities for abuse of power the Founders would *never* have permitted. Now, social and political policies of the entire nation are, through the Fourteenth Amendment and incorporation, vulnerable to manipulation by selective lawsuits, a specialty of the ACLU. We will get a good look in chapter 14 at how that works and how effective it is in manipulating educational policy on a national level.

This situation is intolerable. An amendment to the federal constitution is required to correct this situation and may require nothing more than simply constraining judicial power concerning our enumerated rights to the providence and authority of the individual states. When it comes to our enumerated rights, if the offense is on the part of the state, then deal with it at the level of the state. If the offense is on the part of the federal government, then deal with it at that level.

Further, what the last one hundred years have shown us is that the federal judiciary can no longer be trusted with matters concerning public education, religion, and moral philosophy. While it is proper that protection of the rights of the people under the Bill of Rights should be incumbent on the states, the ultimate expression of those rights and their place in law must be left to the states for this is where the people live, and this is where we find the foundations of the workings of a constitutional republic. Retain *balance-of-powers* and *representative government* and at least we have a chance. The power must remain with the people.

ঙ ঙ ঙ

In *Everson*, the words of the First Amendment had not changed, but somehow they now meant something that you cannot find in the words themselves. This was not a matter of some new principle being "discovered" by an enlightened *Everson* Court that was previously hidden from the Founders. The *Everson* Court simply said in effect that it was time for change and provided the means for subsequent courts to implement that change. Its target? Students in public education. Its purpose? Societal transformation, the covert establishment of a secular religion within government[24] and the neutralization of Christianity as a political and moral force in the American Republic.

Everson opened the door for the religious encroachment the Founders feared, igniting a series of court cases that systematically began to purge any reference to God from public schools. *Engel v. Vitale* (1962)[25] made prayer in public schools unconstitutional, *Abington Township School District v. Schempp* (1963)[26] made reading of the Bible in public schools unconstitutional, *Wallace v. Jaffree* (1985)[27] ruled that public schools could not even observe a moment of silence (not because it was unconstitutional to be silent, but by the very act of setting aside that moment for a religious purpose), and *Lee v. Weisman* (1992)[28] outlawed prayers at public school graduations. Numerous other cases subsequent to *Everson* not only continued the program of purging God from public life but also of rejecting the moral principles that have been foundational to Western civilization for over 2000 years. And no case exhibits this more dramatically than *Roe v. Wade* (1973)[29] and its almost 60 million nameless victims.

❧   ❧   ❧

The decision of the *Everson* Court regarding reimbursement for cost of transportation for Catholic students to and from Catholic schools mattered little to religious liberty. But the precedent established in the Court's misrepresentation of Jefferson's wall of separation and the incorporation under the Fourteenth Amendment of the Establishment Clause radically transformed the judiciary where the federal judiciary assumed for itself jurisdiction over what previously had been the sole domain of the states. I suggest that the motivation, at least in part, of the High Court for this blatant seizure of power might have been nothing less than the politicization of public education to prepare students for a fulfilling life as members of a progressive society guided by positive law. The political payoff is in the future when your adult child, a budding leftist, faithfully votes to continue the process of liberating our nation from the natural law precepts upon which it was founded and for which so many have died.

CHAPTER 14

THE FIRST AMENDMENT AND PUBLIC SCHOOLS

M y original intent for this chapter was simply to review relevant court cases concerning the teaching of origins science in public schools, take those results at face value, and use those results to frame a clear set of guidelines for establishing educational policy that all can agree to irrespective of their religious and political views. By conforming to those guidelines, we should be able to ensure objective science education in public schools and that public education not knowingly undermine the religious beliefs and political preferences of students and their parents. It would seem that a carefully thought-out consideration of First Amendment jurisprudence and scientific principles would provide such a set of guidelines that everyone could be happy with.

It didn't turn out that way. Not at all. Not even close. The problem? Neither First Amendment jurisprudence nor origins science education are being implemented in good faith. Both are corrupted by anti-teleological ideologies. We cannot look to First Amendment jurisprudence in finding a solution to the present problem because First Amendment jurisprudence is part of the problem.

In looking into the legislative and judicial histories of *Everson* and a number of other cases concerning public education, a deeply troubling story began to emerge. It was a story about the encroachment that the Founders feared, an encroachment that today has a steel grip on public education. It is a story about misrepresentation and deception of both science and of judicial principle. For those who still cherish their religious liberties and our founding precepts in natural law, you need to understand this story. Therefore, keeping in mind that I am only a humble engineer and not an attorney or of the academic class, I will tell you that story as best I can. I argue that we cannot hope to untangle and set right the various problems with origins science education in public schools unless we approach that mess with a clear understanding of science, its methods, and its history. Perhaps a true and faithful understanding of science combined with a true and faithful understanding of law will expose the present corruption.

As we saw in *Everson,* an important part of the adjudication process within constitutional jurisprudence entails reference to precedent set by prior courts, generally referred to as *case law*, as well as a concern over the intent of the Founders. Such a practice seems quite appropriate and adds credibility to the work of any given court by anchoring its principle arguments and opinions in an honorable judicial tradition that has deep roots in Western civilization. But there is a problem. That honorable judicial tradition is rooted in *natural law* while today many judges and justices embrace *legal positivism*.[1] What this reduces to is two different kinds of law: one rooted in heaven, the other in the world. We were founded as a nation with laws rooted in heaven. With the subversion of the judiciary by positive

267

law beginning in early twentieth century (or maybe sooner) it is not clear just what principles guide the judiciary today. My simplistic assessment is that liberal jurists tend toward positive law while conservative jurists still retain their roots in the natural law precepts of our founding.

Thus, it is of profound importance for us to remember that the Constitution was written by the Founders within the context of the classical understanding of *natural law*. Accordingly it would be expected that *original meaning* and *intent* could only be understood by reading the Constitution in the same context within which it was written.

In this review and analysis we first set down four judicial principles relevant to the teaching of evolution in public schools. I will call them *JP-1* through *JP-4* to facilitate discussion, provide limited commentary, and then get on with the story. These principles as stated here seem to be completely reasonable; however, the first would not apply during the Founding era, and the second employs reasoning the Founders would likely reject.

JP-1: For First Amendment purposes, *religion* has an *inclusive* meaning that encompasses nontheistic and atheistic as well as theistic systems of belief.

In 1961 the Supreme Court said in *Torcaso v. Watkins:*[2]

> Among religions in this country which do not teach what would generally be considered a belief in the existence of God are Buddhism, Taoism, Ethical Culture, Secular Humanism and others.

In August 2005, the Seventh Circuit Court of Appeals held that *atheism* is a religion. In that case,[3] a Wisconsin prison violated the Establishment Clause when it ruled that a prisoner's desire to create a group to study atheism should be categorized under prison rules as an "activity," not a "religion." The court said:

> In keeping with this idea, [the idea of individual freedom of conscience protected by the First Amendment to select any religious faith or none at all] the Court has adopted a broad definition of 'religion' that includes non-theistic and atheistic beliefs, as well as theistic ones . . . Even if the religion clauses were originally meant only to forestall intolerance between Christian sects, they now encompass all forms of religious Conscience.

This principle accommodates—and rightly so—the diversity of religious views that exist today in America compared to the predominately Christian view of the founding era. Calvert[4] gives a comprehensive review of this subject, providing one of the most important reviews of First Amendment jurisprudence of our times. It is potentially a game-changer.

JP-2: Government must be *neutral* with respect to religion.

In *Epperson v. Arkansas,*[5] Justice Fortas, who delivered the opinion of the Court, said:

> Government in our democracy, state and nation, must be neutral in matters of religious theory, doctrine, and practice. It may not be hostile to any religion or to the advocacy of no-religion; and it may not aid, or foster or promote one religion or religious theory against another or even against the militant opposite. The First Amendment mandates government neutrality between religion and religion, and between religion and non-religion.

In *Rosenberg v. Rector,*[6] in her concurring opinion, Justice O'Conner said:

> The Religion Clauses prohibit the government from favoring religion, but they provide no warrant for discriminating against religion. [citations omitted]. Neutrality, in both form and effect, is one hallmark of the Establishment Clause.

In *Allegheny County v. Greater Pittsburg ACLU,*[7] the court said in ruling against a nativity scene at the entrance to a government office:

> The Establishment clause stands at least for the proposition that when government activities touch on the religious sphere,

they must be secular in purpose, evenhanded in operation, and neutral in primary impact.

In *Gillette v. United States*,[8] the Court said:

> An attack founded on disparate treatment of "religious" claims invokes what is perhaps the central purpose of the Establishment Clause—the purpose of ensuring government neutrality in matters of religion ... Necessarily the constitutional value at issue is "neutrality."

In *Epperson*, the Court also said:

> The State may not adopt programs or practices in its public schools or colleges which 'aid or oppose' any religion. This prohibition is absolute. It forbids alike the preference of a religious doctrine or the prohibition of theory which is deemed antagonistic to a particular dogma.

JP-3: The religious implications of a theory have no bearing under the First Amendment on its permissibility for instruction in public education.

This point was made by Justice Powell in his concurring opinion in *Edwards v. Aguillard:*[9]

> A decision respecting the subject matter to be taught in public schools does not violate the Establishment clause simply

271

because the material to be taught "happens to coincide or har-monize with the tenants of some or all religions."

Thus, while intelligent design has implications that are favorable to theism and while evolution has implications that are favorable to atheism, these implications themselves cannot be used as a basis for forbidding instruction of either subject in public education. The status of intelligent design and evolution as legitimate subjects for *scientific treatment* in public school science education rests on other criteria such as physical evidence, observability, and inductive reasoning, not religious implications. The fact that ID has religious implications does not mean that it is fundamentally a religiously based theory, even if those implications appear to provide or infer an answer to an ultimate question. You could say the same about evolution/neo-Darwinism. The accusation that ID is "religion and not science" does not stand up against court findings.

The *Malnak II*[10] court in Malnak v. Yogi (1977) wrote that:

Certain isolated answers to "ultimate" questions . . . are not necessarily religious answers . . . Thus, the so-called "Big Bang" theory, an astronomical interpretation of the creation of the universe, may be said to answer an "ultimate" question, but it is not, by itself, a "religious" idea.

More from Malnak II:

> A science course may touch on many ultimate concerns,
> but it is unlikely to proffer a systematic series of answers to
> them that might begin to resemble a religion.

Thus, isolated scientific observations and the implications associated with those observations do not constitute the basis for a comprehensive religious belief system subject to the scrutiny of the First Amendment. Yet this precedent is *never* cited in *any* of the relevant court cases concerning the teaching of evolution. Judicial bias was clearly reflected in *Kitzmiller v. Dover* where the defendants (the Dover School Board) had made reference to the atheistic implications of evolution in a newsletter to parents in their district, but the court, instead of acknowledging the simple truth of that observation, used it as evidence to justify the assertion that the members of the school board were religiously motivated.

JP-4: It is permissible to critique evolutionary theory.

It might seem strange that the critique of evolutionary theory would be a concern of the High Court but the *Edwards* Court was careful to point out that its opinion does not . . .

> . . . imply that the legislature could never require that
> scientific critiques of prevailing scientific theories be taught.

Or, in more straightforward language, it is okay to critique prevailing scientific theories—such as evolution. It would have been more appropriate for the Court to say that a state legislature cannot protect a scientific

273

theory from critique just because that theory challenges prevailing religious views.

Since *Everson*, three cases have been heard by the courts concerning the teaching of evolution and creationism in public schools and one case concerning the teaching of evolution and intelligent design. In each, the teaching of creation science or intelligent design was in one way or another found to be in violation of the Establishment Clause of the First Amendment. The overall result was that *evolution-only* is now educational policy for our nation's public schools. The extreme irony here is that the evidence for the theories that have been banned is abundant while the evidence for the theory that is now mandated for instruction in public schools has little or no hard scientific evidence in its support and, given our modern understanding of the biological sciences, *cannot possibly be true*. The courts have either been duped or they are part of the plot against America's children. Maybe both.

∗ ∗ ∗

An examination of these four principles reveals that even within present judicial precedent and case law, there is a legitimate pathway for permitting the teaching of all theories of origins—evolution, creation, and design. But that path is consistently blocked by a judicial majority that is predisposed *against* teaching creationism and design in public schools. While the High Court itself established and/or sanctioned these judicial principles, apparently they are not so sacred that they cannot be sidestepped if a higher purpose demands it. Apparently, banning creationism and design from the public school science classroom is such a purpose.

Take subject matter such as creationism and design in biology. Both have powerful theistic implications, however, under JP-3, those religious implications have no bearing on the permissibility of teaching such subject matter in public schools. Nevertheless, both have been banned from the classroom. So, how did the Court justify banning instruction that judicial principle/precedent (JP-3) says is permissible?

In 1971, the Supreme Court formulated a logical "device" called the Lemon test[11] (to be discussed in some detail shortly) which sets *secular purpose* as the standard for the constitutionality of government action. If there is a question as to the permissibility under the Establishment Clause of some action on the part of government, then it can be subjected to the *Lemon* test to settle the matter. The principle question of concern to *Lemon* was whether or not such action had a clear secular purpose. If so, it is permissible. If not, it is unconstitutional. That seems straightforward.

However, while religious *implication* is permissible under JP-3, it also introduces the possibility of religious motivation, which in turn suggests religious purpose which fails the secular purpose test, resulting in *"death by Lemon"*. Further, criticism of evolution, permissible under JP-4, can also be taken as evidence of religious motivation, also leading to *"death by Lemon"*. However, throw JP-1 concerning the inclusive meaning of religion into the fray and, as we shall see, *Lemon* goes up in smoke, and every single decision beginning with *Epperson* rendered against creationism and design collapses. But, as you might suspect, when it comes to the Establishment Clause, JP-1 is never considered. JP-1 is only considered for cases involving the Free Exercise Clause. This selective application of JP-1 by the Court makes it clear that there is something rotten in Religion Clause jurisprudence.

We will show that in relation to public schools and science education, the underlying *paradigm* within which the federal judiciary functions today is anti-teleological and positivistic in nature, ensuring that even though there may be a legitimate and enduring historical and philosophical connection between God, religion, and the natural sciences, the God and religion connections are banned from the science classroom. This is an abomination to any reasonable concept of justice. Students cannot come to a true understanding of the nature of the natural sciences without understanding the historical and philosophical foundations on which science was conceived.

Much could be written about the subtle perversions of justice on the part of the courts, but I will considered it enough here if we just present the facts and expose the monumental sham. As you will see, that in itself is no small task.

❧   ❧   ❧

In venturing into this area of First Amendment law in relation to science education in public schools, I recognize that this is a highly complex and specialized subject with seemingly endless exceptions, apparent contradictions, and twists and turns through mazes of precedents from which judges and justices can pick and choose in making their arguments and in forming their opinions. Seemingly, if the Supreme Court wants to uphold a statute, it will find a way to do so, or if it wants to overturn that statute, it will find a way to do so. It is just a matter of picking the necessary case law that will lead to the desired end and avoiding case law that might interfere with attaining that end.

I also would like to believe that somewhere in all of this, we can find reasonable and trustworthy common sense judicial principles that can be set down in plain English to guide educational policy in this area of science education that anyone who has taken the trouble to get informed can understand and come to the same logical conclusions that justices of the Supreme Court might reach. In fact, responsibility demands that school board members, state legislators, public education administrators, and even state governors would so inform themselves rather than just place blind faith in staff lawyers or bow to the threats of the ACLU or teachers unions or those science organizations who in the name of science are committed to securing and defending the Darwinian dogma.

The problem, however, with a policy that enforces a high standard of intellectual integrity and objectivity in origins science education is that if there is unscientific dogma present in educational curricula (and there is), then it might be exposed and that cannot be allowed, even if it means *redefining your academic standards in terms of the dogma itself.* Such is the case for Darwinism. The Darwinian dogma must be preserved at all costs. I think we already have an understanding of why.

The commentary that follows does not constitute expert legal opinion, but I believe it does constitute a reasonable description of the state of affairs in both the judiciary and public schools in regard to origins science education. Citations related to the cases of concern are provided for commentary from legal professionals. While those legal commentaries are important to my story, they do not *tell* that story.

Before we take on these four cases, we need to first review the Supreme Court case *Lemon v. Kurtzman* (1971) because of the influence it had on three of these four cases. But even before that, there was an international event in the fall of 1957 that led to a complete reshaping of education in the biological sciences in public schools, and that is what set the stage for the lawsuits to come. I remember this event very well. I was a sophomore in the Corps of Cadets at Texas A&M, Bear Bryant was coach of the Fightin' Texas Aggie football team, John David Crow was pounding defenses like a runaway bull, and I was dating and crazy in love with a pretty little seventeen-year-old girl named Linda. Life was really, really good. It was the fifties. They were happy days. But trouble was brewing.

The event? Sputnik!

✌ ✌ ✌

On October 4, 1957 the Soviet Union launched *Sputnik,* the first man-made earth satellite. This event shocked the world and served to further heighten the intensity of the Cold War that had emerged in the years following the end of World War II. For twenty-one days, Sputnik broadcast its nonstop "beep-beep-beep" (in A-flat) as it orbited the earth, rattling the nerves of free peoples around the world. Then the battery went dead, but the point had been made. The United States' answer to *Sputnik* two months later was dubbed *Kaputnik*, where after the three-stage *Vanguard* rocket had ascended to the dizzying height of four feet, blew up on the launch pad. Under the perception that the Soviet Union was overtaking the United States in science and technology, the National Aeronautics and Space Administration (NASA) was created, and Congress passed the National Defense Education Act to provide funding to develop state-of-the-art science textbooks to reenergize science education in our public schools. The technology race had begun.

In 1958, through a grant from the National Science Foundation, the education committee of the American Institute of Biological Sciences formed the Biological Sciences Curriculum Study (BSCS) with offices in Denver, Colorado. It was through the efforts of the BSCS that evolutionary biology, which had disappeared from biology textbooks for several decades, was reintroduced in those textbooks and the science curriculum of public schools. Darwin had returned to the classroom. *Now, we will show those Ruskies*!

It is of great significance that the BSCS from its inception treated evolution as the central organizing principle of biology, representing it as something more than just a theory about biological diversity. Evolution was given the status of a fundamental principle of the biological sciences, and that principle guided all early BSCS programs. The enthusiasm over neo-Darwinism demonstrated at the Darwinian Centennial in Chicago in 1959 (which in some respects looked more like a religious revival than a scientific meeting) was also reflected in the work of the BSCS. For instance, even as biologists were beginning to recognize the first signs of the coming crisis in evolutionary theory, the *Biology Teachers' Handbook* published by BSCS in 1963 stated "It is no longer possible to give a complete or even a coherent account of living things without the story of evolution." Over the years this *evolution-only* doctrine was institutionalized in public education by writing it into public school biology textbooks, standards and benchmarks, classroom curricula, and assessment tests. What the BSCS actually did was to institutionalize a multibillion-dollar creation myth in public education to counter the Genesis account of the origins of man.

The four cases reviewed here can be correctly understood only in light of this background of the BSCS, its government funding, and its doctrinaire *evolution-only* curriculum. The emergence of the creation science movement in the 1960s was in direct response to this radical Darwinian takeover of education in the biology classrooms of our public schools orchestrated through the efforts and influence of the federally funded BSCS.

Now for *Lemon*.

๑   ๑   ๑

The standard for judicial review in cases involving the Establishment Clause of the First Amendment was established in the 1971 Supreme Court case *Lemon v. Kurtzman*[12] in which the Court set down a test made up of three criteria or *prongs* to determine if a statute or some other government action is permissible under the Establishment Clause. These prongs are generally stated as:

(1) The government's action must have a clear secular purpose;

(2) The government's action must not have the primary effect of either advancing or inhibiting religion; and

(3) The government's action must not result in an "excessive entanglement" of government and religion.

An obvious weakness of the *Lemon* "system" is the inherent subjectivity of the three prongs where there are no criteria set down for what constitutes "a clear secular purpose" or "advancing or inhibiting religion" or "excessive entanglement." One might make a substantial case, especially given the cases to follow, that *Lemon* is self-refuting on the basis of its third prong and maybe even its second prong.

Concern about the possibility of adverse effects on traditional religious views resulting from court rulings based on *secular purpose* were expressed both before and after *Lemon* and with good cause. In the Supreme Court case *Abington Township v. Schempp* (1963)[13] that banned Bible reading and reciting of the Lord's Prayer from public schools, Justice Stewart, dissenting, wrote:

> The state may not establish a "religion of secularism" in the sense of affirmatively opposing or showing hostility to religion, thus "preferring those who believe in no religion over those who do believe." . . . Refusal to permit religious exercises thus is seen, not as the realization of state neutrality, but rather as the establishment of a religion of secularism.

Judge Kiser of the US District Court, in *Crockett v. Sorenson* (1983),[14] in his *Findings of Fact* concerning a course in Bible study (which he found unconstitutional) offered voluntarily to fourth and fifth graders in the City of Bristol, Virginia, school district wrote:

> The First Amendment was never intended to insulate our public institutions from any mention of God, the Bible or religion. When such insulation occurs, another religion, such as secular humanism, is effectively established.

In our modern times, there is a great diversity of worldviews that occupy the place in an individual's life that have traditionally been held by one form or another of theism. Under the *inclusive* meaning of religion, all of those views—whether atheistic, nontheistic or theistic in nature—must, for both Establishment and Free Exercise Clause purposes, be treated as *religion*, forming a comprehensive religious continuum whereby that which *Lemon* calls *secular* is in reality an integral part of that continuum and cannot be clearly defined in any meaningful way as a concept distinct from that continuum. That which *Lemon* calls *secular* is just another brand of religion. Both *Abington* and *Crockett* raised concerns that the *effect* of a ruling of the court that throws out a theory friendly to theistic religion (such as creation science) constitutes the establishment of the religion of secularism. The Supreme Court unambiguously recognizes *secular humanism* as a religion.

We probably should accept that the *Lemon* test was well-intended even though it was ill conceived. When justices and judges form their conclusions regarding the intended *purpose* of government action based on their gut-level perceptions of the *motives* of the sponsors of that action rather than the stated purpose of that action, its logical consequences, or its actual observed effects, then they have entered an intractably subjective domain where you can say goodbye to any standard of objectivity. More on *Lemon* when we look in detail at *Edwards*.

෯ ෯ ෯

The two most important Supreme Court cases concerning the teaching of biological origins in public education were the cases of *Epperson v.*

Arkansas (1968)[15] and *Edwards v. Aguillard* (1987).[16,17] There are many other lower court cases, some before *Edwards* and some after, that also bear on this subject. Of these, the US District Court case *McLean v. Arkansas Board of Education* (1982)[18] was the most notable and important because it was a precursor to *Edwards*. It is in our review of these cases that we come to understand the absolute mess that the Court has made of current judicial principles as they apply to public schools and origins science education.

Epperson v. Arkansas

In *Epperson v. Arkansas* (1968), the Supreme Court overthrew a 1929 Arkansas statute enacted by the Arkansas legislature as a consequence of the 1925 Scopes trial in Dayton, Tennessee, that prohibited the teaching of *human* evolution in all state-supported educational institutions, including universities and public schools. The Arkansas statute stated "It shall be unlawful . . . to teach the theory or doctrine that mankind ascended or descended from a lower order of animals." This restriction applied to both teachers and textbooks.

In 1965, Susan Epperson, a tenth-grade biology teacher, was given a new textbook for instruction which contained a chapter about evolution, the product of the work of the BSCS. This (supposedly) created a dilemma for Susan in that if she complied with the instruction she (supposedly) would be subject to dismissal under the 1929 statute and if she didn't, she could likewise be subject (supposedly) to dismissal for insubordination. There is also good reason—make that *very* good reason—to believe that the whole affair was concocted by the ACLU. She filed suit in the

Chancery court, seeking a declaration that the 1929 statute was void and she should be granted protection from dismissal by the state if she used the subject textbook. The court ruled in her favor, holding that the 1929 statute violated *freedom of speech* and *academic freedom*. On appeal, the Arkansas Supreme Court reversed the decision, holding that the state of Arkansas had the power to specify curriculum in public schools.

On further appeal, the *Epperson* Court held that the states could not use the First Amendment to justify tailoring teaching and learning to the requirements of any religious sect or dogma, and in particular, the teaching of the evolutionary origins of mankind could not be prohibited by the state just because it was contrary to the belief that man was created by God. While Justice Black concurred with the opinion of the Court, he foresaw that the failure of the courts to recognize that evolution conflicts with the religious beliefs of many Americans would cause much public controversy over this issue in the years to come. In his concurring opinion he wrote[19]:

A second question that arises for me is whether the Court's decision forbidding the State to exclude the subject of evolution from its schools infringes the religious freedom of those who consider evolution an anti-religious doctrine. *If the theory is considered anti-religious, as the Court indicates, how can the State be bound by the Federal Constitution to permit its teachers to advocate such an "anti-religious" doctrine to schoolchildren? . . . Since there is no indication that the literal Biblical doctrine of the origin of man is included in the curriculum of Arkansas schools, does not the removal of the subject of evolution leave the State in a neutral position toward these*

supposedly competing religious and anti-religious doctrines?
Unless this Court is prepared to just write off as pure nonsense
the views of those who consider evolution an anti-religious doctrine, then *this issue presents problems under the Establishment
Clause far more troublesome than are discussed in the Court's
opinion.* (my emphasis)

The concerns raised by Justice Black are just as valid today as they
were in 1968. He understood that the *Epperson* Court, in handing down
its opinion, in effect endorsed a materialistic religious view of the origins
of mankind, contradicting the view embraced by the Founders and held by
the great majority of Americans even today that our existence is the result
of the creative work of God, not chance material processes.

⋘ ⋘ ⋘

In public education, the concept of *neutrality with respect to religion*
was first mentioned in *Everson* in the words "Our public school ...
[should] maintain a strict and lofty neutrality as to religion" but was set
down as judicial doctrine in *Epperson. Epperson* said:

> Government in our democracy, state and nation, must be
> neutral in matters of religious theory, doctrine, and practice.
> It may not be hostile to any religion or to the advocacy of
> no-religion; and it may not aid, or foster or promote one
> religion or religious theory against another or even against the
> militant opposite. The First Amendment mandates government

neutrality between religion and religion, and between religion
and non-religion . . .

While *Epperson* formalized the doctrine of *neutrality with respect to religion* in Establishment Clause jurisprudence, that concept itself was brought into question by none other than the late Chief Justice William Rehnquist. The chief justice addressed this question in his dissent against the majority opinion of *Wallace v. Jaffree*[20] (1985) in which he provided much-needed historical context as to the relationship between government and religion. His argument was that the Founders did not believe that government should be *neutral* toward religion at all, and that this was reflected in the history of those times, both with respect to the debates concerning the First Amendment as well as the legislative actions of the First Congress. He also cites the opinions of the two leading constitutional scholars of the nineteenth century, Joseph Story and Thomas Cooley. Story was a justice of the United States Supreme Court from 1811 to 1845, professor at the Harvard Law School, and author of *Commentaries on the Constitution of the United States*. Cooley was chief justice of the Michigan Supreme Court and dean of the University of Michigan Law School. He also authored a major treatise on constitutional law first published in 1868.

The focus of Chief Justice Rehnquist's dissent in *Wallace* was the false and deceptive metaphor fabricated by the *Everson* Court of Thomas Jefferson's "wall of separation" and the consequences of its incorporation by *Everson* into Establishment Clause jurisprudence. And since one of the rotten fruits of *Everson* was the *Lemon* test, he includes *Lemon* in his critique as well.

First, we summarize the religious atmosphere of the founding era and then see how that atmosphere was reflected in Chief Justice Rehnquist's critique.

∽ ∽ ∽

The Founders asked the question, "What form of government is best suited to ensure the safety and happiness of the citizens of this new nation?" The answer to that question rested (as described in chapter 12) on the understanding of the nature of man and the relationships between God, man, and government as prescribed in the Declaration of Independence. The Founders were convinced that a *republic* was the only form of government under which man could enjoy both civil and religious liberty. They also understood that the republic they were creating could not endure unless virtue, Christian virtue, prevailed among the people and their elected leaders.

Accordingly there was an inherent *affability* of the Founders, even the non-Christian Founders like Jefferson and Franklin, toward Christian faith because they knew that without Christian virtue among the people rooted in an understanding of the life and teachings of Jesus, the Republic could not endure. The framework and the laws of the new nation would be spelled out on parchment, but virtue must be written on the heart and passed on through the generations by the people. Christian virtue was the *silent partner* in establishing the Republic, and the survival of the nation literally depended on it. It still does.

Some have referred to the *press* as "the Fourth Estate" in its job as watchdog over government. In like manner, Christian virtue must be acknowledged as "the Zeroth Estate" in providing the moral foundation of the Republic. The government that was established in late eighteenth century was positive in its general support to Christianity but cautious in ensuring that no favoritism was shown among Christianity's many denominations. Ultimately, it was up to the people as to whether they would propagate the Christian faith and the virtues it commanded.

These ideals are clearly reflected in the history of our country, and it is that history to which Chief Justice Rehnquist appealed. In his dissent, the chief justice reviewed the events of late summer and early fall of 1789 when the First Congress adopted the language for the First Amendment, passed the Northwest Ordinance, and adopted a joint resolution asking President George Washington to issue a Thanksgiving Day Proclamation. Brief excerpts from the chief justice's dissent give us a sense of the prevailing mood relative to religion. It also establishes a clear context for the meaning and intent of the Religion Clause of the First Amendment put in place during the founding era and as it persisted throughout the nineteenth century.

Concerning the Northwest Ordinance:

> The Northwest Ordinance ... provided that "[r]eligion, morality, and knowledge, being necessary to good government and the happiness of mankind, schools and the means of education shall forever be encouraged."

Concerning the Thanksgiving Proclamation:

The Presidential Proclamation said: "Now, therefore, I do recommend and assign Thursday, the 26th day of November next, to be devoted by the people of these States to the service of that great and glorious Being who is the beneficent author of all the good . . . that we may then all unite in rendering unto Him our sincere and humble thanks for His kind care and protection . . . for the signal and manifold mercies and the favorable interpositions of His providence . . . for the great degree of tranquility, union, and plenty which we have since enjoyed . . . for the civil and religious liberty with which we are blessed . . . for all the great and various favors which He has been pleased to confer upon us . . . that we may then unite in most humbly offering our prayers and supplications to the great Lord and Ruler of Nations, and beseech Him to pardon our national and other transgressions . . ."

Concerning the general mood of Congress:

As the United States moved from the 18th into the 19th century, Congress appropriated time and again public moneys in support of sectarian Indian education carried on by religious organizations. Typical of these was Jefferson's treaty with the Kaskaskia Indians, which provided annual cash support for the Tribe's Roman Catholic priest and church saying in part '*And whereas,* the greater part of said Tribe have been baptized and received into the Catholic church, to which they are much attached, the United States will give annually for seven years one hundred dollars towards the support of a priest of that religion ... [a]nd ... three hundred dollars, to assist the said Tribe in the erection of a church.' It was not until 1897, when aid to sectarian education for Indians had reached $500,000 annually, that Congress decided thereafter to cease appropriating money for education in sectarian schools."

Commentary by Joseph Story as quoted by Chief Justice Rehnquist:

Probably at the time of the adoption of the Constitution, and of the amendment to it now under consideration [First Amendment], the general if not the universal sentiment in America was, that Christianity ought to receive encouragement from the State so far as was not incompatible with the private rights of conscience and the freedom of religious worship. An attempt to level all religions, and to make it a matter of state policy to hold all in utter indifference, would have created universal disapprobation, if not universal indignation.

The real object of the [First] [A]mendment was not to countenance, much less to advance, Mahometanism, or Judaism, or infidelity, by prostrating Christianity; but to exclude all rivalry among Christian sects, and to prevent any national ecclesiastical establishment which should give to a hierarchy the exclusive patronage of the national government. It thus cut off the means of religious persecution (the vice and pest of former ages), and of the subversion of the rights of conscience in matters of religion, which had been trampled upon almost from the days of the Apostles to the present age.

Commentary by Thomas Cooley as quoted by Chief Justice Rehnquist:

But while thus careful to establish, protect, and defend religious freedom and equality, the American constitutions contain no provisions which prohibit the authorities from such solemn recognition of a superintending Providence in public transactions and exercises as the general religious sentiment of mankind inspires, and as seems meet and proper in finite and dependent beings. Whatever may be the shades of religious belief, all must acknowledge the fitness of recognizing in important human affairs the superintending care and control of the Great Governor of the Universe, and of acknowledging with thanksgiving his boundless favors, or bowing in contrition when visited with the penalties of his broken laws. No principle of constitutional law is violated when thanksgiving or fast days are appointed; when chaplains are designated for the army and navy; when legislative sessions are opened with prayer or the reading of the Scriptures, or when religious teaching is encouraged by a general exemption of the houses of religious worship from taxation for the support of State government. Undoubtedly the spirit of the Constitution will require, in all these cases, that care be taken to avoid discrimination in favor of or against any one religious denomination or sect; but the power to do any of these things does not become unconstitutional simply because of its susceptibility to abuse.

[t]his public recognition of religious worship, however, is not based entirely, perhaps not even mainly, upon a sense of what is due to the Supreme Being himself as the author of all good and of all law; but the same reasons of state policy which induce the government to aid institutions of charity and seminaries of instruction will incline it also to foster religious worship and religious institutions, as conservators of the public morals and valuable, if not indispensable, assistants to the preservation of the public order.

Within the context reflected in the excerpts above, it is hard to pull any meaning from the Establishment Clause of the First Amendment beyond that noted by Chief Justice Rehnquist:

One would have to say that the First Amendment Establishment Clause should be read no more broadly than to prevent the establishment of a national religion or the governmental preference of one religious sect over another.

According to the chief justice, the Establishment Clause means what it says, nothing more and nothing less. In late eighteenth century, any attempt to interpret it otherwise would likely have been met with sound resistance. Here are some excerpts from Chief Justice Rehnquist's dissent that summarize his view of *neutrality*:

The Establishment Clause did not require government neutrality between religion and irreligion nor did it prohibit the Federal Government from providing nondiscriminatory aid to religion. There is simply no historical foundation for the proposition that the Framers intended to build the "wall of separation" that was constitutionalized in *Everson.*

Whether due to its lack of historical support or its practical unworkability, the *Everson* "wall" has proved all but useless as a guide to sound constitutional adjudication. It illustrates only too well the wisdom of Benjamin Cardozo's observation that "Metaphors in law are to be narrowly watched, for starting as devices to liberate thought, they end often by enslaving it."

But the greatest injury of the "wall" notion is its mischievous diversion of judges from the actual intentions of the drafters of the Bill of Rights. The crucible of litigation . . . is well adapted to adjudicating factual disputes on the basis of testimony presented in court, but no amount of repetition of historical errors in judicial opinions can make the errors true. The "wall of separation between church and State" is a metaphor based on bad history, a metaphor which has proved useless as a guide to judging. It should be frankly and explicitly abandoned.

Thomas Cooley noted that the general affability of the federal government toward religion was expressed toward Christianity alone. There was no hint of the idea of *neutrality* with respect to Christianity to be found during the founding era or throughout the nineteenth century in America. *Neutrality with respect to religion* in general is solely a construct of the post-*Everson* era.

McLean v. Arkansas

In *McLean v. Arkansas* (1982), a Federal District Court declared as unconstitutional Arkansas Act 590 of 1981 called the "Balanced Treatment of Creation-Science and Evolution-Science Act." The essence of this act was that no teacher was required to teach either creation science or evolution science, but if one was taught, both had to be taught. The language of the act gave broad latitude to the schools and teachers on how this educational balance was to be achieved and specifically restricted instruction to scientific evidence only, prohibiting religious instruction or references to explicit religious writings.

Act 590 had its pedigree in a "balanced treatment model" developed by the Institute for Creation Research (ICR) for use at the level of school boards but was adapted by Paul Ellwanger of Anderson, South Carolina, in developing a *model act* that he promoted to state legislatures. Arkansas was the first state to adopt the *model act*. Louisiana was the second (and last) to adopt it. Lawsuits were filed in both Arkansas and Louisiana. The act provided details in what was meant by creation-science and evolution-science to guide classroom instruction:

(a) "Creation-science" means the scientific evidences for creation and inferences from those scientific evidences. Creation-science includes the scientific evidences and related inferences that indicate: (1) Sudden creation of the universe, energy, and life from nothing; (2) The insufficiency of mutation and natural selection in bringing about development of all living kinds from a single organism; (3) Changes only within fixed limits of originally created kinds of plants and animals; (4) Separate ancestry for man and apes; (5) Explanation of the earth's geology by catastrophism, including the occurrence of a worldwide flood; and (6) A relatively recent inception of the earth and living kinds.

(b) "Evolution-science" means the scientific evidences for evolution and inferences from those scientific inferences and related inferences that indicate: (1) Emergence of naturalistic processes of the universe from disordered matter and emergence of life from nonlife; (2) The sufficiency of mutation and natural selection in bringing about development of present living kinds from simple earlier kinds; (3) Emergence by mutation and natural selection of present living kinds from simple earlier kinds; (4) Emergence of man from a common ancestor with apes; (5) Explanation of the earth's geology and the evolutionary sequence by uniformitarianism; and (6) An inception several billion years ago of the earth and somewhat later of life.

Ellwanger's model was enacted by the Arkansas legislature with little ado and signed into law by Governor White on March 19, 1981. On May 27, 1981, a suit was filed in US District Court challenging the constitutionality of the act on the grounds that it was in violation of the Establishment Clause of the First Amendment.

To provide proper historical context for the "Balanced Treatment Act," it should be noted that by the time this act was introduced in the Arkansas legislature in 1980, twenty years of federal funding had been invested in the BSCS in developing curriculum and textbooks (Judge Overton noted this in his Memorandum Opinion), in effect institutionalizing Darwinism in public schools. No such investment had been made by the federal government in creation science. Thus, there was this prevailing institutional bias against the idea of teaching creation science, which seemed to many parents, scientists, educators, and legislators to be *out of balance*, especially given the favor shown to evolution in public education by refusing to expose students to its deep and pervasive evidentiary problems that raised doubts as to its veracity even within mainstream science.

Plaintiffs were made up of a diverse lot that included various mainstream Protestant and Catholic church officials and clergy, parents and friends of minor children attending Arkansas Public Schools, a number of Jewish organizations, the Arkansas Education Association, the National Association of Biology Teachers, and the National Coalition for Public Education and Religious Liberty. The ACLU provided legal representation for the plaintiffs. Defendants included the Arkansas Board of Education and its members, the Director of the Department of Education, and the State Textbooks and Instructional Materials Selecting Committee.

McLean was quite an event. In 1982, Darwinism was a theory in crisis as reflected by the scientific dissent from Darwinism taking place at that time within mainstream science. Darwinist sensitivities were at an all-time high. Their theory and their materialistic faith were in trouble.

The Arkansas act posed a serious threat to the evolution-only goal of Darwinism, and *McLean* was the last line of defense. An illustrious list of evolution's big guns, such as Michael Ruse, Steven J. Gould, George Marsden, Harold Horowitz, and Francisco Ayala testified. Judge Overton dutifully followed the lead of the esteemed witnesses and of course the ACLU. If one carefully reads Judge Overton's Memorandum Opinion and distills it all down to the central argument that is brought to bear in finding the Arkansas act unconstitutional, it is his construction of demarcation criteria for distinguishing between what is, and what is not, science. He then subjects creation science to this criteria wherein he concludes that creation science does not qualify as *science*, and because it is not science it must be *religion*, and because it is religion, the framers of the act must be religiously motivated. If they are religiously motivated, then the purpose of the act is to advance a religious view, thereby failing *Lemon's* first prong, which requires a clear secular purpose. The contrived nature of the reasoning cobbled together by Judge Overton was deeply disturbing to creationists.

But the creationists were not the only ones disturbed by Judge Overton's reasoning. Mainstream philosophers of science, though delighted with the overall outcome, were also deeply disturbed by Judge Overton's demarcation criteria and the contrived means by which he arrived at his conclusions. They were further troubled by the concern that if you want to win an important court case for the home team (evolution), you would like that victory to be based on something *substantive*, not contrived. From their view, *McLean* was a victory that rang hollow.

Philosopher of Science Michael Ruse was a key expert witness for the prosecution in *McLean*. Years later (1996), Ruse pulled together a collection of essays on this case in his book, *But is it Science?*[21] (referring to creation science). Ruse admits that the purpose of this collection (in which he authors the prologue and eight of the twenty-seven essays that make up the collection) was to defend his own views against his critics. To his credit, he gives everyone a free shot at him, and they all fire away. Overall, it makes a valuable contribution to the continuing debate between creationism and evolution and gives rare insight into the thinking of leading materialistic philosophers of science on this subject. This collection is particularly enriched by including important historical works by William Paley, Adam Sedgewick, and Thomas Huxley and a paper on creation by the venerable Duane Gish. Finally, while there was nothing balanced about Judge Overton's opinion of the Balanced Treatment Act, the collection of essays Ruse put together was actually quite balanced. It would be nice if public education was granted the liberty to have an open debate about our origins, following the example of Michael Ruse who invited such a debate and published the results.

It is interesting that in these essays, especially in the writings of Larry Laudan, a highly regarded philosopher of science, you get the impression that intellectual integrity is of far greater concern to philosophers of science than it is to Darwinian scientists. Laudan specifically condemned Judge Overton's use of demarcation criteria where he says that philosophers of science do not use such criteria anymore. His view of what happened in Arkansas is summarized in his essay, "Science at the Bar,"[22] where he says:

> The victory in the Arkansas case was hollow, for it was
> achieved only at the expense of perpetuating and canonizing a
> false stereotype of what science is and how it works. If it goes

unchallenged by the scientific community, it will raise grave
doubts about the community's intellectual integrity.

It has been thirty-five years, and we are still waiting for that challenge
from the scientific community but so far, not a peep.

Judge Overton issued his opinion on January 5, 1982, but *McLean* was
never appealed, not because creationists were willing to give up without a
fight but because by that time Louisiana ACT No. 685, a similar act with
the same ICR pedigree, was being challenged. Some important changes
had been made to the Louisiana act, which were expected to reduce its
vulnerability to a challenge in the courts; thus, the creationists shifted
their efforts to the Louisiana battle that would eventually wind up at the
Supreme Court in *Edwards v. Aguillard*.

When Senator Bill Keith first introduced his bill in the Louisiana
legislature in 1980, its content was somewhat different than that of the act
that was sent to Louisiana Governor Edwards in 1981 for signature. The
purpose of the bill in its original form was to "assure academic freedom
by requiring the teaching of the theory of creation ex nihilo in all public
schools where the theory of evolution is taught" and defined the theory of
creation ex nihilo as "the belief that the origin of the elements, the galaxy,
the solar system, of life, of all the species of plants and animals, the origin
of man, and the origin of all things and their processes and relationships
were created ex nihilo and fixed by God." In the post-*Everson* world, this
choice of words and the entanglement of creationist doctrine with science
as part of the purpose of the act was problematic. Unfortunately, even
though this language was not retained in the final version of the act, the

act itself could not be separated from its legislative history, and it was that history that would eventually contribute to its downfall.

The lawsuit concerning the Arkansas act was filed on May 27, 1981. Upon learning of this action, Senator Keith, submitted a revised draft of the Louisiana bill the very next day, stripped down to its bare essentials with the hope that this minimalist version might avoid a challenge in the courts such as had occurred in Arkansas.

The Louisiana act was passed overwhelmingly and signed into law by Governor Edwards in July 1981. Immediately, Louisiana educators (Department of Education, Superintendent of Education and the Board of Elementary and Secondary Education) rebelled, refusing to implement the act because they were convinced that its real purpose was to teach creationism. It is interesting that the *educators* defended the act in Arkansas but challenged the act in Louisiana.

A lawsuit immediately erupted, which bounced around in the judicial system for five years as summary judgements, dismissals, stays, and appeals, involving the US District Court in Baton Rouge (1981), the US District Court in New Orleans (1981), the Louisiana Supreme Court (1983), the US District Courts again (1985), the Fifth Circuit Court of Appeals (1985), and finally the US Supreme Court (1986) where the ruling by the Fifth Circuit was appealed.

The stated purpose of the simplified bill was to "protect academic freedom," and while the meanings of creation science and evolution science remained unchanged, the detailed lists (1) through (6) of what constitutes evidence for creation science and evolution science found in the Arkansas act were not included in the Louisiana act. They could not, however, escape the legislative history of the bill, and it was that history, whether justified or not, that provided the basis for a religious *motive*. Once religious motive was "established," it no longer mattered what the language was or the sincerity of the sponsors of the act in their good-faith attempts to separate the legitimate scientific and educational content of the act from the Genesis account. Under the original language of the act, there would be little difficulty in making the case that what creation science really boiled down to was simply dressing up the Genesis story in the language of science. Even though "balanced treatment" *really was* the good-faith purpose of the Louisiana legislature, this undeniable connection to Genesis became the principle rationale for taking creation science out of public education for good. From the day it was passed to the decision in *Edwards*, *"death by Lemon"* was preordained for the Louisiana bill.

Edwards v. Aguillard

In *Edwards v. Aguillard* (1987), a 7–2 majority struck down the Louisiana Balanced Treatment of Creation-Science and Evolution-Science Act. The *Edwards Court* in deciding the constitutionality of the Louisiana Balanced Treatment Act applied the first prong of the *Lemon* test to the act, which required that the act have a clear secular purpose. The Court found that the act failed that test because it had a *religious* purpose rather than a *secular* purpose. Even though the act explicitly stated that its purpose was to "protect academic freedom," a perfectly legitimate assertion with a clear secular purpose, Justice Brennan rejected that assertion as a sham because in looking at the history of the legislation and earlier versions of the act, he concluded that the framers of the act were religiously moti-vated, and because they were religiously motivated the true purpose of the act, regardless of its actual language and good-faith intent, must be to advance religion. In reaching this decision, the *Edwards Court* happily ignored reams of evidence, including the testimony of experts who had been tasked by the Louisiana legislature to examine the legitimacy of a secular purpose in teaching a curriculum that included creation science. But it was to no avail. Justice Brennan was hell-bent on overturning the Balanced Treatment Act, which was not adjudicated on anything substantive, such as its content and stated purpose or any reasonable expectation of what its effect might be, but rather on the subjective perceptions of the justices of the motives of the framers of the act. Francis Beckwith[23] provides an extensive review of Justice Brennan's majority opinion as well as a summary of Justice Scalia's scathing dissent.

What happened in *Edwards* was appalling. The Louisiana legislature, having learned from what happened in Arkansas, had written an act that was, both on its face and in its substance, scientifically, educationally, and constitutionally sound. The conceptual leap of the act from *balanced treatment* (which was what the act provided) to *academic freedom* (which was the act's purpose) was perfectly reasonable. There is little doubt that the framers of the act did in fact entertain some level of religious motivation, after all they were Christians, and many saw the teaching of evolution as antagonistic to Christian faith. Given the years of discrimination and marginalization of creationists and their views by the scientific and educational establishments, it seemed quite reasonable that a statute be enacted that would protect this group. Finally, it is no small matter for the High Court to overturn such a statute, but for the High Court to accuse such a body of duly elected representatives of the people of *sham legislation* is inexcusable. Given the circumstances, we have far better reason to trust in the integrity of the Louisiana legislature than the *Edwards* majority. This is what judicial tyranny looks like and this is how it acts.

It is impossible to escape the conclusion that in going after the religious motivations of the framers of the Louisiana act, it was the *Edwards* majority, not the Louisiana legislature that was religiously motivated. Common sense would suggest that just as the religious *implications* of a theory of science are incidental to the constitutionality of that theory (JP-3), so likewise should the religious *motivations* of the framers of such legislation be incidental to the constitutionality of that legislation. Legislation should be judged on its constitutional merits, not subjective, perceived motivations of the legislature. Regardless, unable to find the act unconstitutional on its merits, the Balanced Treatment Act was subjected to "*death by Lemon*".

The enormity of the blow to religious liberty and the offense to the constitution realized through a single ACLU lawsuit are almost incomprehensible. One of the most cherished rights of the people within a constitutional republic is that laws are enacted by representatives of the people, not the nine justices of the Supreme Court.

The net effect of *Edwards* was the establishment of a secular religion in public education, not just in Louisiana, but (thanks to the Fourteenth Amendment) in all fifty states in the union. That evolution is in reality a *secular* religion was frankly admitted by Michael Ruse, former Christian and self-proclaimed Darwinist. In his article (2000) in the Canadian newspaper *National Post*[24] Ruse said:

> Evolution is promoted by its practitioners as more than mere science. Evolution is promulgated as an ideology, a

secular religion—a full-fledged alternative to Christianity, with meaning and morality. I am an ardent evolutionist and an ex-Christian, but I must admit that in this one complaint—and Mr. Gish [referring to the late creationists Duane Gish] is but one of many to make it—the literalists are absolutely right. Evolution is a religion. This was true of evolution in the beginning, and it is true of evolution still today.

Evolution therefore came into being as a kind of secular ideology, an explicit substitute for Christianity.

⊰ ⊰ ⊰

The consequence of the *Edwards* decision reached far beyond the legitimate domain of the judiciary and public education. In reaching its decision, the *Edwards* Court betrayed the trust of the people it serves and violated its sacred mission of protecting the natural rights of man. Ever since *Everson,* the overreach beyond "action" to "opinion" as Jefferson put it, has been freely violated by the courts. The *Edwards* Court, in condemning the religious motivations of the framers of the Balanced Treatment Act and by mandating that public schools teach the fundamentals of a naturalistic religion rooted in the natural origins of mankind, did all that the Establishment Clause forbade and did it in the *name* of the Establishment Clause. The *Edwards* majority not only extended its reach beyond *action* to *opinion*, it then brought government action *against* opinion where the basis of that judgement was its *own* opinion. What ever happened to principle?

The Louisiana legislature attempted in good faith to restore objectivity and balance to origins science education. The Supreme Court not only said "no", it stripped *all of the states*, not just Louisiana, of the power to protect the families they serve from government encroachment against the foundational doctrine of theistic religion, belief in a Creator. Further, this action on the part of the *Edwards* Court violated the central principle of a constitutional republic, the right of the people to govern themselves.

The responsibility for making state and federal laws rests with state and federal legislators elected by the people to represent the interest of the people. The phrase "due process of law" pertains to the protections under law afforded to the people and guaranteed under the Fifth and Fourteenth Amendments. The process of *establishing* law is hard-wired into federal and state constitutions and carried out by elected representatives of the people. In the spirit of the Fifth and Fourteenth Amendments, we might call this "due process of representation" meaning the right of the people through representation to make the laws under which they live. "Due process of law" concerns the rights of the people to the protections provided under law. "Due process of representation" concerns the rights of the people to participate in the law-making process, or more to the point, the right to *self-government*. *Edwards,* through the Fourteenth Amendment, violated "due process of representation", thus trampling on the core precept of a constitutional republic, the right to self-government. This process has been repeated over and over again since 1947.

But there is more.

◈   ◈   ◈

To illustrate the sheer arrogance and hypocrisy of the *Edwards* decision, we conduct a simple exercise whereby we subject the *Edwards Court* and its decision to the same judicial standard to which the Louisiana act was subjected: the *Lemon* test. It would be completely legitimate, quite practical, and even praiseworthy for the *Edwards Court* itself, in its concern for the integrity of its decision to invite critique by submitting its own decision to the *Lemon* test. Further, and of greater importance, would be the objective of ensuring that the *remedy* imposed by the Court does not impose a greater evil than the *problem* it supposedly cured. Surely the Court in the process of righting one wrong would not want to impose a greater wrong, especially on all fifty states within the republic.

Let us apply the *Lemon* test to the decision of the *Edwards* Court. If the Supreme Court is going to hold an act of a state legislature to a particular judicial and constitutional standard, then we should expect that the decision of the Court itself should stand up to those same standards. After all, the Supreme Court is no less subject to the US Constitution than a state legislature. Consequently, it makes sense that the High Court be held to the same standard in its decisions that it exercises in its judgements. Sounds like a jolly good idea to me, one that should be made into law and imposed as standard procedure, especially in those cases that touch on matters such as our religious liberties. Maybe the Congress would like to think about that, and maybe *the people* would like to give their elected representatives their opinion on the subject. This would provide at least some accountability of the High Court by holding in check the corrupting influences of the power they hold.

On with the *Lemon* test. First, we give *Edwards* a "pass" on *Lemon's* first prong regarding secular purpose because of its inherent subjectivity, which is more than *Edwards* did for the Louisiana act. Accordingly, we go straight to the second prong which requires that the primary effect of the act must *not* be to *advance* or *inhibit* religion, criteria which are far more amenable to an objective assessment than the first prong. What we find is that for the last thirty years, the theory of evolution as a naturalistic account of the origins of mankind (which is the central dogma of all naturalistic religions such as atheism, naturalism, materialism, secularism, and humanism) has been taught exclusively in public education as undisputed fact and the teaching of creation science (which is the central dogma of all monotheistic religions, such as Christianity, Judaism, and Islam) has been banned. Thus we see that *Edwards* violates the second prong of *Lemon* and does so on two counts, one by advancing naturalism and the other by inhibiting Christianity, and it has been doing this without interruption for the last thirty years. Further, for each school year that passes, this decision touches the lives of roughly 50 million students, and it will do so throughout their K-12 years in public schools.

> *Edwards, in using the first prong of Lemon to throw out the Louisiana Balanced Treatment Act, failed the second prong, miserably.*

Edwards was a direct assault on our religious liberties and the Constitution of the United States. It is unthinkable that the *Edwards* decision should be permitted to stand.

A simple consideration by the *Edwards* Court of the necessity of incorporating an *inclusive* meaning of religion in adjudicating the constitutionality of the Louisiana Balanced Treatment Act would have led to a completely different result, one that, as will be shown shortly, turns *Lemon* upside down. But that could not be allowed because *Edwards* was about vengeance, not justice. You could "feel it" in the tone of Justice Brennan's majority opinion. In the end, was *Edwards* simply a "hate crime"?

Kitzmiller v. Dover Area School District

In December 2005, Federal District Judge John Jones III delivered his findings in the *Kitzmiller v. Dover Area School District* lawsuit,[25] which was the first court case to directly consider the constitutionality of teaching intelligent design in public education. The charges before the court were that the Dover School Board acted for religious purposes when it issued its science policy to be read to ninth-grade biology students, which included the statement:

> *Intelligent Design is an explanation of the origin of life that differs from Darwin's view. The reference book, Of Pandas and People, is available for students who might be interested in gaining an understanding of what Intelligent Design actually involves.*

First, it should be said that neither intelligent design nor Darwin have anything to do with specific theories about the origin of life. Such are the perils of what happens when school boards, no matter how well intended, set out on their own to write science education policy that deals with scientific content, especially in the area of origins science, whether it be the origin of the universe, the origins of life, the origins of species, the origins of mankind, or the origins of mind and consciousness. Regardless, the Dover Science Policy statement had no influence on curricular materials, and there was no requirement for taking any classroom time for the subject of intelligent design other than the three-minutes it took to read the policy statement. If there is one single lesson that can be taken from this case, it is that within the scientific, educational, and judicial establishments, there can be no toleration whatsoever of even a hint of a challenge to Darwinism or the secular dogma that dominates public education. Any uprising, no matter how insignificant, must be crushed. The Dover Science Policy was an easy target, providing one more judicial victory to honor the rotten corpse of Darwinism. For the scientific establishment the problem of containing evidence of God's creative work in nature has reduced to a game of "whack-a-mole." A "whack" in Pennsylvania, and peace was restored in Darwin-land all across America. At least for the moment.

The presumptuous nature of Judge Jones' findings is jaw-dropping. He "found" that ID is not science, evolution is not contrary to belief in a divine Creator, and ID arose from creationism rather than a scientific dissent from Darwin. He simply believed what the ACLU told him. The good judge was completely duped.

❧ ❧ ❧

It is a long-established principle of constitutional jurisprudence that in any given case, the court should focus on the issues that have been brought before the court and should not venture further than necessary to render a decision. This principle was blatantly violated in *Everson*, and it was violated in *Kitzmiller* as well. Justice Black allowed reimbursement of the cost of public transportation for Catholic students and then moved on to much bigger fish, rewriting Establishment Clause jurisprudence. Judge Jones could not top that, but as a Federal District judge, he did all right for himself. Rather than simply rule on the constitutionality of two sentences in the Dover Science Policy, he ran amok putting the entire intelligent design movement on trial. There was no compelling judicial reason for either Justice Black's or Judge Jones' actions. With respect to Judge Jones, however, there was every appearance that he was on a mission, and that mission was to drive a stake through the heart of intelligent design. He understood the historic nature of this case, saw it as Scopes II, and he was the presiding judge. His place in history was secured, and in the eyes of the scientific establishment, Judge Jones was a rock star.

≼ ≼ ≼

In a way, the whole Kitzmiller affair was rather sad. Judge Jones had no clue as to the true history of the ID movement and its connection to the scientific dissent from Darwinism. One reporter said that in an interview, Judge Jones said that he was going to review the film *Inherit the Wind* in preparation for the trial. At that point, there was little doubt as to the outcome.

The short book *Traipsing into Evolution*,[26] authored by Law Professor David DeWolf, et al., summarizes major concerns with Judge Jones' findings in *Kitzmiller* and served as a guide to what I have included here. Some of DeWolf's concerns are as follows:

(a) In his findings, Judge Jones determined that members of the School Board were religiously motivated, and it was on this basis that he declared the science policy to be unconstitutional and thus subject to *"death by Lemon"*;

(b) Methodological naturalism was central to the argument for discrediting intelligent design as science;

(c) Judge Jones concluded that intelligent design was "an interesting theological argument, but that it is not science, and because it was not science, the only effect that could result was to advance religion (I can see philosopher of science Larry Laudan[27] rolling his eyes and shaking his head);

(d) Even a cursory examination of the findings in Kitzmiller would show that all four of the judicial principles discussed at the beginning of this chapter were not just ignored, they were violated;

(e) Judge Jones grossly misrepresented the testimony of Professor Michael Behe where it was obvious that he (Jones) had simply performed a "cut and paste" on substantial material from the ACLU's proposed "Findings of Fact and Conclusions of Law" submitted by the ACLU a month before the ruling; and

(f) Judge Jones, without regard to its scientific merits or relevant legal precedence, erroneously concluded that because intelligent design has theistic religious implications, it is *religion* and not *science*. So much for JP-3.

❧ ❧ ❧

In the first decade of the twenty-first century, intelligent design was becoming a serious threat to the evolution-only policies in public education throughout America. Darwinists and the ACLU were eager for a court case to eliminate this threat just as in 1987, *creation science* was eliminated by the *Edwards* Court. The Dover Board of Education, naïve as to the nature of the legal mine field they were walking into, provided the perfect agenda for such a case. Judge John Jones, the perfect judge, was easily manipulated by the ACLU. Thus, good intentions of the Board of Education notwithstanding, *Kitzmiller* was "The Perfect Storm" in its destruction of the principles of First Amendment jurisprudence, objective scientific inquiry, and religious liberty in public schools. One could hardly orchestrate a greater disaster for Establishment Clause jurisprudence.

Finally (this is personal for me), the mistreatment and abuse of Professor Behe whose scientific work and personal integrity stand on their own, was *disgusting* testifying to the fact that this trial had little to do with science, law, and intellectual integrity but much to do with the wielding of institutional power for the cause of secular dogma and political power.

Americans need to understand the extent of the judicial and scientific corruption at work in *Kitzmiller*. As law, *Kitzmiller,* under the manipulation of the ACLU, was an abomination. It may very well be that the outcome of *Kitzmiller* played out in many of the federal district courts across the country would have been the same, though we would like to think that most would at least address the issues before the court, rather than use the court to seek fame.

The ACLU and the scientific and educational mafias played Judge John Jones III like a banjo. I hope someday he will come to understand that. I also hope that for his own sake, he will have an opportunity for redemption in whatever form it may take. He's not the bad guy here. He just got conned by the best con-artists on the planet. Great damage was done by *Kitzmiller* to public education, religious liberty, scientific integrity, and the integrity of the judiciary.

The Problem with *Everson*

We have already exposed the deception in the majority opinion written by Justice Hugo Black in his stunning and blatant misrepresentation of Jefferson's "wall of separation" and its meaning and the impact it had on succeeding courts. Ever since *Everson*, the mandate of "neutrality with respect to religion" seems to be the accepted standard within the judiciary and admittedly on its face it seems reasonable, almost *self-evident* you might say. But not really. The judges and justices of our courts seem to care very much about the religious attitudes and opinions of the Founders and often looked back to that era for understanding. However, when we look at what the Founders actually said about "neutrality with respect to religion," we find that they said nothing. There was nothing "neutral" about the religious spirit of the founding era, even for those who did not profess Christianity. In fact, if the colonists had been neutral with respect to religion, there would never have been a Revolutionary War and an American Republic.

Justice William O. Douglas reflected positively on the role of religion in American life when in the 1952 case of *Zorach v. Clauson*[28] he wrote:

> "When the state encourages religious instruction . . . it follows the best of our traditions. For it then respects the religious nature of our people and accommodates the public service to their spiritual needs. To hold that it may not would be to find in the Constitution a requirement that the government show a callous indifference to religious groups. That would be preferring those who believe in no religion over those who do believe . . . We find no constitutional requirement which makes it necessary for government to be hostile to religion.

Then in 1961 in *McGowan et al. v. Maryland*,[29] fourteen years after *Everson*, Justice Douglas said:

> The institutions of our society are founded on the belief that there is an authority higher than the authority of the State; that there is a moral law which the State is powerless to alter; that the individual possesses rights, conferred by the Creator, which government must respect. The Declaration of Independence stated the now familiar theme: 'We hold these Truths to be self-evident, that all Men are created equal, that they are endowed by their Creator with certain unalienable Rights, that among these are Life, Liberty, and the Pursuit of Happiness.' . . . And the body of the Constitution as well as the Bill of Rights enshrined those principles . . . The Puritan

influence helped shape our constitutional law and our common law as Dean Pound has said: The Puritan "put individual con- science and individual judgment in the first place." For these reasons we stated in *Zorach* v. *Clauson*, "We are a religious people whose institutions presuppose a Supreme Being."

A reading of the history of the founding era does not reflect in any form or to any degree whatsoever the neutrality of *Everson* and *Epperson*, much less the secularism of *McLean, Edwards*, and *Kitzmiller.* This concern was raised by Chief Justice Rehnquist in *Wallace v. Jaffree* in his dissent and elaborations (summarized earlier in this chapter) on the misuse by the Court of the metaphor of Jefferson's "wall of separation." Justice Black's misrepresentation of Jefferson's "wall" and "neutrality with respect to religion" are not legitimate constitutional principles concerning the rela- tionship between government and religion. Rather, within an inclusive meaning of religion, *secular* is as much an anti-Christian viewpoint as atheism or humanism.

In light of the monumental principles and events entailed in the founding and history of the American Republic, it would seem that the judges and justices of our judiciary would have an ever-present spirit of humility and gratitude considering the sacredness of their office and the high calling of protecting these treasured principles. To render justice is doing the Lord's work. Since *Everson*, however, it seems that much of the work of the judiciary has been dedicated to the eradication of such a spirit from both the judiciary and from public life.

As Chief Justice Rehnquist explained in Wallace[30] (with lavish para- phrase and elaboration on my part), "It is egregious error to assume that by excluding a specific subject from the classroom under an order of the Court that you teach nothing about that subject. On the contrary, what you teach to children is that this subject has explicitly been determined by the authority of the highest court in the land and the ruling scientific and educational establishments so that it is not just a matter of secondary importance or of no importance at all, but worse. It is excluded, cast out, accursed, shunned, rejected, scorned, and banished to the outer darkness, reserved for such nonsense and the fools that embrace it. Within main-stream science, a curse will fall on any who speak of such a thing with favor" (lavish paraphrasing and elaboration concluded).

Thus, in public education today, creation science and intelligent design are still being "taught," not openly, which would invite legitimate critique and rousing discussion, but silently, invisibly, insidiously. The message: they are not worthy of mention. They are to be forgotten.

This is indoctrination, it is imposed by law on every student in public education in America throughout their formative years, it suffocates learning, it starves students of a true understanding of the extraordinary natural world in which they live, and it prepares them for life in an oppressive progressive state.

The Problem with *Epperson*

Epperson (1968) was decided three years before *Lemon* (1971), and thus the options offered by the *Lemon* test were not available to the *Epperson* Court in reaching their decision. The *Epperson* Court held that the states could not use the First Amendment to justify tailoring teaching and learning to the requirements of any religious sect or dogma and in particular that the teaching of the evolutionary origins of mankind could not be prohibited by the state just because it was contrary to the belief that man was created by God. *Epperson* was decided through use of an *exclusive* meaning of religion, which leads to a simple creed with broad application: *creation by God is* "religion" and *creation by evolution is* "science." This was the logical basis for deciding *Epperson* and no doubt strongly influenced *McLean, Edwards* and *Kitzmiller* as well.

However, under an *inclusive* meaning of religion, you could just as well say that the states could not use the First Amendment to justify tailoring teaching and learning to the requirements of any religious sect or dogma and in particular that the teaching of the theory of creation of man by God could not be prohibited by the state just because it was contrary to the belief that man was of evolutionary origins. Touché, Justice Fortas!

Now the problem within the federal judiciary is that in precisely those situations in which the *inclusive* meaning of religion is necessitated, it is ignored in favor of the traditional exclusive (theistic) meaning. Consequently, when the *Epperson* decision was decided, based on an *exclusive* theistic meaning of religion, the net effect was to grant to evolution a favored and protected status within both American jurisprudence and public education and to ban theories supporting belief in the divine creation of man.

If public education is going to take it upon itself to guide students in the study of the natural world by means of the natural sciences, then under the *inclusive* meaning of religion, it must be prepared to accept the fact that the same evidence for a guiding intelligence that has been acknowledged throughout much of recorded history might very well be observed and understood as such by students in public education today.

317

Further, neither the courts nor public education have the constitutional power, even under the mandate of methodological naturalism, to deliberately misinform students in regard to scientific evidence concerning the origins or nature of man. When methodological naturalism is constrained to operate under an *inclusive* meaning of religion, its function as a deceptive naturalistic "device" is fully exposed. You might say it is "castrated." Under the inclusive meaning of religion public education cannot shape educational curriculum to "fit" *any* religious account of the origins of mankind, including evolutionary biology, which under the terminology of Michael Ruse, is a "secular religion."

Lemon and the Problems with *McLean, Edwards,* and *Kitzmiller*

Having seen the crucial role of the *Lemon* test in *McLean, Edwards,* and *Kitzmiller,* we now take a closer look at *Lemon* for the purpose of gaining a better understanding of its underlying assumptions and of determining whether or not those assumptions are appropriate to the cases (other than *Lemon*) in which the *Lemon* test was applied.

In 1971, the Supreme Court, in *Lemon v. Kurtzman,* ruled on the constitutionality of education statutes in Rhode Island and Pennsylvania, both of which involved a substantial administrative interaction between state government and Catholic schools within their states. The Rhode Island statute provided a 15 percent salary supplement to be paid to teachers in those nonpublic schools in which the average per-pupil expenditure on secular education was below the average in public schools. Eligible teachers were restricted to teaching only courses offered in the public schools, using only materials used in the public schools and had to agree not to teach courses in religion. In Pennsylvania, the statute authorized the state to purchase secular educational services from nonpublic religious schools, directly reimbursing those schools for teachers' salaries, textbooks, and instructional materials. The great majority of those nonpublic schools in both Rhode Island and Pennsylvania were Catholic schools, and many of the teachers in those schools were nuns.

A lower court had found the Rhode Island statute to be in violation of the Establishment Clause, while in Pennsylvania, a lower court had dismissed the complaint against the Pennsylvania statute on a technicality. Both cases were heard on appeal by the Supreme Court.

In both Rhode Island and Pennsylvania, the statutes were well-intended with regard to secular education but state-sponsored teaching of secular subjects in nonpublic religious schools unavoidably brought church and state interests into close proximity in two respects. First, it was necessary for the state to impose requirements for extensive monitoring of educational records and bookkeeping along with classroom surveillance to ensure that classroom instruction in church schools supported with state funds was carried out without religious content. Second, grave concern was expressed by the *Lemon* Court about the ability of teachers in religious schools (especially nuns) to avoid introducing religious content into instruction in secular courses sponsored by the state. Normally, in religious schools, the mixing of religious and secular content would be routine and unremarkable because that is the way it works in real life.

In the course of the *Lemon* proceedings, the Court developed the *Lemon* test with its three prongs to provide guidelines for its adjudication of the constitutionality of the two statutes. In both cases the Court found that the statutes satisfied the first two prongs (secular purpose and neither advance nor inhibit religion) but failed the third on the basis of *excessive entanglement* between religion and state. It is not clear how the Court was able to equate *excessive entanglement* to a violation of the Establishment Clause. Given the provisions of the statutes, avoiding excessive entanglement between religion and state would certainly be prudent but if that entanglement did not involve any action that was plainly in violation of the Establishment Clause or the Free Exercise Clause (and there is no evidence at all that it did), then it is hard to understand how you can get "unconstitutional" out of "excessive entanglement." It was likely that it was the concern over inadvertent mixing of religious and secular content that triggered the *excessive entanglement* prong. Regardless, this is not our concern.

◈ ◈ ◈

As to whether a case should be decided using an *exclusive* as opposed to an *inclusive* meaning of religion turns on the specific conditions that pertain to the question brought before the courts. *Lemon* was decided on the basis of an *exclusive* meaning of religion in that it involved only Catholic schools, secular education, and state funding. No questions concerning religious doctrine were raised by the statutes, the *exclusive* (theistic) meaning of religion was both implicit and appropriate, and within the context of the Rhode Island and Pennsylvania statutes, there was nothing about the meaning of "secular" as it pertained to subject matter like literature, mathematics, economics, and history that would be offensive to Catholicism. Thus, under the *exclusive* meaning of religion, the use of "secular" in the first prong of *Lemon* was entirely appropriate;

it embodied no inherent antagonism toward theistic faith and ensured neutrality with respect to religion, a precedent established by *Epperson* in 1968. Within the context of the two statutes, the overlap between state and religion was in fact quite "soft" and had to do only with the mechanics of providing secular education in religious schools.

The *Lemon* test "fit" the *Lemon* case very well because it was designed for the *Lemon* case. It did not, however, "fit" *McLean, Edwards,* and *Kitzmiller* at all because it was not designed to address the specific conditions that pertained to the Arkansas and Louisiana statutes or the Dover Science Policy.

The conditions concerning *McLean, Edwards,* and *Kitzmiller* were dramatically different from those of *Lemon.* These cases specifically touched on fundamental dogma concerning two competing theories of the origins and nature of mankind, one naturalistic and the other teleological. On one side was evolution, which provided a naturalistic account of the origins of mankind. On the other side was creation (*McLean* and *Edwards*) with obvious ties to the Genesis creation account, and intelligent design (*Kitzmiller*) with its reference to intelligent agency. A long, controversial history dating back to the Scopes Trial in 1925 and the publication of Darwin's *Origin of Species* in 1859 preceded *McLean, Edwards,* and *Kitzmiller,* a history that challenged fundamental Christian beliefs and shook the foundations of Western civilization. By contrast, the thing that was remarkable about *Lemon* had nothing to do with *Lemon* itself or the opinion of the *Lemon* Court. Rather it was the influence it had on *McLean, Edwards,* and *Kitzmiller* through the application by those courts of the *Lemon* test.

Because evolution gives an account of the origins and nature of mankind, it is inherently religious in nature. It is the *nature* of the question itself that determines the philosophical or religious status of theories that are brought to bear on that question. Even if you come up with a secular answer to a religious question, that answer is inherently religious in nature because the word *secular* is religious in nature. Our multicultural and religiously *inclusive* society demands an *inclusive* meaning of religion, and if the rights guaranteed under the First Amendment are to be secured, then this *inclusive* nature of religion must also be secured in First Amendment jurisprudence.

Under the *inclusive* meaning of religion necessitated by the specific conditions introduced by the Arkansas and Louisiana statutes and the Dover Science Policy, everything changes. "Secular" now aligns philosophically with such godless religions as naturalism, materialism, humanism, and atheism, where all in varying degrees are antagonistic to belief in God and theistic faith. The *Lemon* test was inappropriate to *McLean, Edwards,* and *Kitzmiller* because the adjudication of those cases all necessitated an *inclusive* meaning of religion.

Under the *inclusive* meaning of religion, the first prong of the *Lemon* machinery collapses because it is impossible to distinguish in any meaningful way between *secularism* and *naturalism* or any other of its naturalistic kin. "Secular" in the first prong of *Lemon* subjected to an *inclusive* meaning of religion could just as well read "must have a clear naturalistic, materialistic, humanistic, or atheistic purpose." If the federal judiciary has a lick of integrity left, it will take action to correct these grievous offenses committed against the children of America.

જી   જી   જી

We now see the importance of the work of legal scholar John Calvert.[31] His analysis is centered on *Kitzmiller*, but it applies generally and is foundational to the arguments presented here. The key to ensuring the integrity of science education in public schools and in protecting the rights of parents to direct the religious education of their children without subversive interference by the state, is the mandate that scientific theories about the origins of mankind be taught objectively without religious or political bias, that educators and judiciary alike respect the distinction between the religious implications of a theory of science and the theory itself (JP-3), and that Establishment Clause constraints be imposed solely within an *inclusive* meaning of religion (JP-1).

Considering the extent and depth to which Darwinism is entrenched in public education, cleaning up this mess will require drastic measures. There is no reason why biological evolution should not be taught in our public schools but only if the draconian practice of the last thirty years of protecting evolutionary theory from legitimate scientific critique, defending it as undisputed scientific fact and protecting it from competing theories, must end.

જી   જી   જી

These reviews of *Everson, Epperson, Lemon, McLean, Edwards*, and *Kitzmiller* expose the *deliberate* and *deceptive* abuse on the part of the judiciary whereby the constitutional power of judicial review has been used as opportunity to effect social, cultural, and political change. The concept of neutrality with respect to religion and the requirement for a predominant secular purpose in government actions are more than dubious doctrines. They are false doctrines. There is not a trace of such thinking in the founding era or for the first 100 years of the American Republic. In our own times, longstanding precedent rooted in natural law has been thrown out, and new precedent based on misrepresentations of original constitutional meanings and intent have been institutionalized under the influence of a positivist political and legal philosophy. What the last seventy years have taught us is that because of the positivist law and progressive political doctrines that have infiltrated the federal judiciary, it can no longer be trusted with matters concerning religious faith, moral philosophy, and the education of our children.

The judges of the courts and justices of the Supreme Court, are highly qualified scholars and practitioners at *law*, but they are not qualified to provide truly *independent objective review* of the complex and contentious arguments that arise concerning science education that touches on the origins of mankind and ultimate matters of religious faith. So long as they see these conflicts as part of a "war" between science and religion, their opinions will be biased against religion, especially Christianity and in particular, conservative evangelical Christianity, the brand of Christianity that launched the American Revolution and which refuses to yield to Darwinism.

Roe v. Wade

Regardless of the view you may hold concerning abortion, anyone who cares about *due process of law*, *due process of representation* and judicial integrity should be deeply troubled, not only by the opinion handed down in *Roe v. Wade*,[32] but by the deception that undergirds that opinion. It is a little-known fact that this landmark case was decided within the context of a completely fabricated and false account of 800 years of abortion law in England and America (much in the same way that *Everson* enshrined a completely fabricated and false account of Thomas Jefferson's "wall of separation"). The new orthodoxy concerning abortion law established by Justice Harry Blackmun who wrote the majority opinion, entails four theses: (1) that abortion has always been a common practice in human history; (2) that voluntary early abortions were not crimes until the nineteenth century; (3) that nineteenth-century abortion statutes were designed to protect the life of the mother rather than the life of the child; and (4) that abortion statutes were enacted on behalf of a male-dominated medical profession to eliminate competition from female midwives. The findings in *Roe v. Wade*, the legal justification for overthrowing existing abortion laws in all fifty states, the "discovery" of a "right" to privacy in the Constitution that encompasses abortion, and the abortion of 60,000,000 unborn babies over the last forty-five years, all rest on this false history. Without this concocted history, the Court would never have been able to fabricate what might pass as a "legal" right to abortion. And herein we find a fundamental principle of positive law: If you want to concoct a new "right," first concoct a false history that reflects a distorted view of reality that embodies that "right," and then concoct a lawsuit designed to affirm the constitutionality of that "right." Under positive law you are not encumbered with things like eternal truths (the createdness of all things) or moral principles (the sanctity of life) but rather are free to construct legal principle to serve the needs of the circumstances (sex without consequences). That is the whole

325

point of positive law. Whatever deception is required is of no concern because the end justifies the means. Using a concocted nonexistent history, the *Roe* majority concocted a non-existent right to abortion out of thin air.

The massive deception in *Roe v. Wade* is all chronicled by Joseph W. Dellapenna, professor of law at Villanova University, in his almost 1300-page volume, *Dispelling the Myths of Abortion History*,[33] a work fifteen years in the making. While it might be noted that Villanova is a Catholic university, Dellapenna is not Catholic but identifies himself as a "backslidden Unitarian" and confesses that he has no particular opinion concerning abortion one way or the other. He does, however, seem to have quite a strong opinion concerning judicial integrity. Dellapenna's tome is surprisingly readable. He knows this fascinating story very well, and he knows how to tell it. In excruciating detail, Dellapenna takes us through the actual history of abortion law in England and America, especially during the past 200 years, plainly showing that abortion laws were centered on protection of the life of the unborn child. This was the basis of abortion law in all fifty states, all of which were overturned by *Roe*.

≼ ≼ ≼

The fraudulent history upon which the Court relied was concocted by Cyril Means Jr., general counsel for the National Association for the Repeal of Abortion Laws (NARAL). Fully half of Justice Blackmun's majority opinion was devoted to that fraudulent history where it was noted that copies of Means' articles "documenting" that history were observed on the bench with the justices during oral arguments. Without Means' bogus history, the only rational outcome of *Roe* would have been that it would simply re-validate existing abortion law in all fifty states. In fact, without Means and his "history", there never would have been a lawsuit. A heavy question hangs in the air. How could the justices of the *Roe* Court *not* know that the history of abortion law presented in testimony before the Court was bogus?

The historical and philosophical significance of what the High Court did in *Roe* can properly be understood only in light of the natural law principles of the Declaration of Independence, the *first things*. One may pose arguments on the basis of biology as to what constitutes life, or look to medical science in determining when life begins. And one may postulate legal arguments as to whether the life of the unborn child is protected under the law. But the overwhelming consensus of 800 years of law was that the protection of the life of the unborn child came first.

As for what our founding precepts have to say about life, it is quite simple. Life is a given, a fact of existence, and it is sacred. God is our Creator and gives us life and has endowed us with our natural rights, our rights to life and liberty. Under the Constitution, all of law and every aspect of government must conform with the *first things* of the Declaration—at least that was the way it was before Darwin and *positivism* came to America.

A created order is a divine order. If that is in fact true, then there are some places within that order where man cannot go, some things he cannot touch, some things he cannot look upon, and some things that are sacred. The majority in *Roe* went where it had no right to go, touched sacred things it had no right to touch, and the consequences were horrible. The *Roe* majority not only erred, it sinned. Some sins are so grievous, so fundamental that they can only be described as a sin against nature. Such was the case for *Roe v. Wade*.

Surprisingly, *Roe* has also come under criticism from pro-choice legal analysts who fault the Court for its failure to note that when it comes to abortion, there is far more at stake than privacy. A *life* is at stake. Further, they fault the Court (and rightly so) for usurping the role of the state legislatures in crafting abortion law, where it is likely, given the prevailing mood of the American people at the time, that most states would have nothing to do with such a thing. Regardless, when it comes to matters of such fundamental importance as the nature of man and the sanctity of life, *never again* must the Supreme Court be allowed to overturn the laws of all fifty states with a single opinion.

◆ ◆ ◆

The proceedings of the *Roe* Court were conducted in cold, sterile, clinical terms where the justification for taking the life of an unborn child was reduced to historical, medical, and legal considerations. The purpose was to dodge the deeper truth that in abortion, a heart stops beating, a human life is extinguished, and there are unavoidably moral consequences. Having accepted the false history of abortion law, the *Roe* majority was able to find a "right-to-privacy" implicit in the Fourth Amendment which encompassed the constitutional right for a mother to give up the life of her unborn child, thus illustrating the power of positive law which has an almost unlimited capacity to recast problems with actual moral concerns into a cold uncaring framework devoid of even a trace of morality. In *Roe,* when it came to a choice between the privacy rights of the mother and the life of the unborn child, the child had neither a say nor an advocate.

Thus, in *Roe,* we see a head-on collision of *positive law* with *natural law*. Under natural law, all life is sacred. Under positive law, nothing is sacred. Unanchored from natural law, positive law is free to respond to the needs of the situation unencumbered by things such as universal, eternal, unchanging Truths. Justice Blackmun had a rich history of case law available to cite which was devoted to protecting the life of the unborn child. However, he had an agenda and that agenda had nothing to do with the natural law precepts of our founding.

∾ ∾ ∾

It is impossible to miss the ideological connection between Darwinism and abortion as well as their relationship in positive law. If life and mankind emerged as a result of unguided natural processes alone, then from the perspective of the Darwinian fundamentalist, it is hard to argue that there is any essential difference between a human and a mosquito. Abortion or mosquito whacking—in the end, they are the same thing. But somehow, deep in our souls, we *know* that they are not the same thing because we know there is something fundamentally different between a mosquito and an unborn human child.

However, when it comes to abortion, only the most radical materialists believe that it is as simple as whacking a mosquito. It is a traumatic ordeal for a mother to decide to end the life that is within her, regardless of the circumstances. There is simply no way to trivialize such a thing. But this is where Darwin comes in. There is a balm for the emotional aftermath of abortion, not in Gilead, but in Darwin.

> ***Darwin broke the connection between mankind
> and the Divine.***

At least that is the "official" story. That is the way the devout see it and that is the accepted view of the academy. What Darwin actually did, however, was to construct a rather ingenious hypothesis that allows us to *imagine* that this connection might be broken—or at least it might be loosened a bit. That enables us to entertain the possibility that maybe God "is not watching" after all, or maybe He is not watching as closely as we had imagined. Darwinism gives us a rationale for arguing that maybe abortion is not such an awful sin against God after all, or maybe it is not a sin at all. In fact, maybe in reality, Darwin liberated us from the burden of the stringent biblical moral code that has enslaved much of Western civilization for most of its history. So maybe, in that sense, abortion is a good thing.

Or maybe not.

> ***Intelligent design reestablished the connection between
> mankind and the Divine that Darwin supposedly broke.***

While in the twentieth century, Darwin may have broken the connection between mankind and the divine, by early twenty-first century, *intelligent design had firmly reestablished that connection,* and that contributes in part to the insane intolerance of the scientific establishment and progressive political establishment for the scientific aspects of design.

And whatever we might have thought about that connection between man and God being severed or at least eased a bit by Darwin, the natural sciences tell a different story, a story about a universe in which mankind is central to its design and purpose. If that is true then it must also be true that procreation and unborn children have a central place in this design and purpose as well. Maybe there *is* a divine plan here, one even the Supreme Court should respect and take seriously.

The connection between man and God was not broken by science. Rather, the natural sciences have affirmed that connection and today the light of that connection shines brightly on *Roe* and all its dark places.

॰ॐ ॰ॐ ॰ॐ

In the twentieth century, the doctrines of materialism and positivist epistemology spread throughout all of our public institutions, corrupting everything they touched—the natural sciences, political philosophy, moral philosophy, judicial philosophy, and public education. The *Edwards* and the *Roe* Courts together recast fundamental questions as to the nature of man (*Edwards*) and the sanctity of human life (*Roe*) in terms of a concocted science (evolutionary biology), a concocted motivation for government action (secular purpose), a concocted history (abortion law), a concocted law (right to privacy), and a concocted right (right to abortion), all concocted to sidestep the central issues concerning our understanding of the true nature of man and the sanctity of life. Never mind for the moment our Christian convictions and legacy. Simply considering who we are as a nation and as a people (who admittedly cannot be separated from their faith), what should be our point of reference in our understanding of the

331

nature of man and the sanctity of life? Under positive law, the nature of man and the sanctity of life are irrelevant, having been reduced to a set of concoctions. But in the American Republic, that point of reference is, always has been, and always will be the natural law principles of the Declaration of Independence, the *first things* revealed by nature. The natural sciences, as shown herein, rather than contradict these principles, affirm the thinking of the Founders concerning the place of man in the universe. Man is of divine origins, and life is sacred. That changes *everything*.

∾ ∾ ∾

The central question for Christian parents in America today is whether their children will reach maturity with sufficient learning, wisdom, and foundation to be able to see through and to withstand the misrepresentations and deceptions that have been institutionalized within our culture by the scientific, educational, political, media, and even the judicial establishments. There is a whole "other" world out there committed to making sure our children never discover who they really are and what their lives really mean, that they never come to understand their true place in either the universe or the nation in which they live. Under the present assault, what hope is there that your child will reach maturity with his or her love of God and Country still intact?

That "other" world intends to prepare them for life in a society that the Founders would absolutely reject, one in which our individual and corporate obedience and devotion in this life is offered to the secular state rather than to God.

CHAPTER 15

SUMMARY

T his descent into the "Matrix" described in the Preface, has revealed that far from living in a meaningless world governed by chance, as the scientific establishment would have us believe, we live in an amazing universe, filled with hope and meaning beyond our wildest dreams. There is an ultimate reality, an ultimate truth, that can be known. Mankind is not an accident of nature; he is central to the meaning and purpose of the universe. The American Founders in their wisdom knew that this was true and built a nation on that truth.

However, these great truths about the nature of man and the founding precepts of our nation are under assault from forces that now dominate most of the institutions of our culture and in particular, public education. The purpose of this treatise has been to identify and help us come to an understanding of some of these unsuspecting but crucial aspects of our culture that are used to control our lives, not in the abstract but within the context of our own times, our nation, our history, our faith, the natural sciences, our understanding of the natural world, our understanding of the nature of man and what our lives mean.

Unfortunately, in America, public education has become a very big part of that world. In that world, we are taught that there are no fundamental truths, that one way is as good as another, and that there are no real moral choices to make, merely *options* from which to choose. It is a world with many "paths". We are told that they are all "right," and we are free to choose our own way, but the truth is that most lead to destruction. How can we find our way in such a world? What will be our guide? Who can we believe? Who can we trust?

Ultimately, there can be only one Truth. This Truth comes to us through both divine revelation and nature, and it is *that* Truth upon which all else turns. *That* Truth as revealed to us by the natural sciences is that mankind is central to the design and purpose of the universe. Our existence is the result of the creative work of a designing purposeful intelligence. Through the natural sciences we have seen the "fingerprints of God" all over His creation. The early Christian theologians were right. There *are* two books—the book of God's Word and the book of God's Works—and these books tell two sides of the same story, a story about creation and a story about redemption, stories that are very good news for mankind.

These same truths were also reflected in the natural law precepts of the Declaration of Independence and the founding of the American Republic. The Founders understood that we have our existence in a created order, man is God's favored creation, and our rights to life and liberty are given by God. No man and no institution of man can stand between a man and his God. The sacred mission of government is to honor and protect—not govern and regulate—the fundamental rights of man in his quest to know, serve, and worship God. And if mankind is in fact central to the design and purpose of the cosmos, then every other fact of existence must be understood and all knowledge and learning of our children sanctioned, under and in relation to that central Truth. That was the vision of the founding generation, and that was the America that Alexis de Tocqueville[1] saw when he visited America in 1831. If we as a *people* cannot unite around these founding principles and renew our commitment to God and country according to those original precepts, then it is inevitable that we will forfeit the blessings of liberty. The threat is from *within*, and it is relentless.

Given our understanding of the biological and cosmological sciences today, we are left with the conclusion that naturalism, materialism, atheism, secularism, and humanism are irrational systems of belief. They may accord with a *positivist* version of reality constructed by men, but they do not accord with an *objective* reality revealed by nature. The essential dogmas of these belief systems—the natural origins of the universe and the natural origins of life, of consciousness and mind, and of mankind—go against the fundamental ontological message of the natural world. They do not "fit" reality as revealed to us through the natural sciences. Contrary to the dismal message of the scientific establishment, the heavens really do declare the glory of God.

❧ ❧ ❧

I will now undertake to summarize the main points of this treatise. Thank you for hanging in there to the end.

THE WARFARE MYTHOLOGY

The so-called *war between science and religion* is a myth, a myth crucial to maintaining a general oppressive atmosphere that protects the teaching of evolution in public education, accommodates methodological naturalism in censoring evidence of design and purpose in the natural world from public education, and conditions the mindset of jurists. It ensures that regardless of the question that might arise concerning the conflict between Darwin and design, there will be a preexisting and overwhelming favorable predisposition toward Darwin, regardless of the merits of the evidence. All arguments are reduced to "Darwin is science and *design* is religion".

In reality, however, there *is* a war, but it is not between science and religion. It is between religion and naturalism where science and religion are partners joined *together* in eternal conflict against their common enemies, Darwinism and naturalism.

DARWINISM

Darwin postulated that life progressed through a long, gradual, cumulative process of small changes from simple forms leading to the diversity of living creatures and eventually the emergence of mankind. However, our modern understanding of biology and cosmology tells us that Darwin was far more than just wrong. He got it backwards. The environment did not shape life, much less create it. First there was the universe, then a place called Earth, then life, then mankind.

> *Mankind is the culmination of a creative process that had him in mind.*

It is a universal truism that the potter shapes the clay, not the other way around. Darwinists say that matter came first, and mind emerged from matter. But that is not what the natural world tells us. As Professor Wald found,[2] mind came first. In fact *mind* is eternal, and *matter* came from *mind*. That is the only rational interpretation of what we observe in nature. It really does not matter that materialistic scientists, for materialistic reasons, reject that general account of history. The unmistakable message of the cosmological and biological sciences today is that the universe was purposely designed for man as opposed to the idea that man was accidentally "designed" by the universe.

Darwinism came to maturity under the positivist philosophy of science, which was formalized by Ernst Mach and the Vienna circle in the early twentieth century. The only antidote for Darwinism, as is true for all positivist theories, is scientific truth, and such truth can be found in abundance throughout nature but *not* in the halls of science or public school biology classrooms. The evolutionary account of the history of life and the natural origins of mankind is a gigantic institutionalized *scam*. This scam touches on matters that reach far beyond the education of our children to their indoctrination into a view of the world that rejects the natural law precepts of our founding.

337

> ***Darwinism is a theory that fails to explain something
> that never happened.***

Public education really should not, in the name of science, indoctrinate students so that they will believe, along with the Red Queen of Wonderland, impossible things, especially impossible things that are contrary to the prevailing religious faith of the American people and the natural law founding of the American Republic.

THE DARK SIDE

It is hard to comprehend the depth and extent of the misery that was brought down on mankind in the twentieth century by evil, tyrannical political systems that employed *eugenics* as social policy in their efforts to establish their utopian empires. It is even harder to comprehend that American universities funded by wealthy American progressives would take part in engineering that misery. We cannot cast blame directly on Darwin for this, but he was certainly aware of the dark implications of his theory for mankind, and there is little doubt that these things deeply troubled this kindly old gentleman during his later years.

In the late nineteenth and early twentieth centuries, scientific understanding of biology was simply insufficient to judge conclusively one way or the other as to the legitimacy of Darwin's theory, never mind eugenics. Consequently, its authority was not established under the *inductive* paradigm as in the "hard" sciences, such as thermodynamics and classical physics. Rather it was established as a *deduction* from materialism through a system of "just-so" stories, and it was under the authority of these "stories" that its place was secured in secular scientific culture and then imposed on the broader culture through public education and social policies. The American scientific establishment of early twentieth century took an unsubstantiated, highly improbable hypothesis and used that hypothesis to justify their creation of a monster they eagerly set loose in America and Germany. The result was some of the darkest times in the history of mankind when evil ruled over much of the earth.

If students are to be taught about Darwin's theory of evolution and the origin of species based on materialistic "just-so" stories, then public education is obligated to inform these students about the horrors, atrocities, and misery that were poured out on mankind in the name of Darwinism as well. All of that suffering, all of that horror, all of that terror, all of that blood, all of that slaughter over some "just-so" stories. There is far more to the story of evolution than students are being taught in public schools. They are entitled to hear about "the rest of the story"—the dark side. In fact, they cannot possibly understand the last 100 years of our history and our modern times if they do not understand that story. They should understand that eugenics and the misery and suffering it wrought cannot be laid solely at the feet of Darwin, and neither can it be laid at the feet of science. It is the responsibility of scientists, mere humans who in their arrogance presumed that they were qualified to judge what constituted "defective" human beings. There is a lesson here and it has nothing to do with biology.

THE NATURAL SCIENCES

Modern science arose in Christian Europe in the sixteenth and seventeenth centuries, there and nowhere else. It was the rational Christian doctrine of a knowable created order that made this possible. No other philosophical or religious system of thought expressed a worldview that provided the unique mindset necessary for science to emerge.

Bacon's observational, experimental, and inductive methodological tradition provided the rational framework for the development of science. This tradition led to Newton's theory of mechanics, Maxwell's theory of electromagnetism, and Einstein's space-time physics, together known as classical physics. The astonishing success of classical physics testifies to the legitimacy of the underlying teleological premises of the natural sciences that came together 500 years ago in Christian Europe. In early twentieth century, however, under the influence of materialism and a positivist epistemology, we descended into the *Dark Ages* of science.

Today we are told that there are two distinct domains of physics—classical and quantum. The reason that there are two domains is not that there really are two domains; rather, we have simply created those domains in our minds. Our thinking "ran off the rails." We adopted two distinct theories of knowledge—realism and positivism—based on two distinct views of the world—teleological and materialistic. The teleological/real system based on *law and design* gave us classical physics and the foundation of the natural sciences. The materialistic/positivist system based on *law and chance* gave us Darwinism and quantum mechanics.

There is only one world and that world is *real* and *classical* and it is governed, not by *law* and *chance,* but by *law* and *design.* If science is to ever find its way out of the *Dark Ages*, it must abandon the materialistic/ positivist world of *law and luck* and return to realism, classical physics, and the methodological tradition of Bacon.

METHODOLOGICAL NATURALISM

Methodological naturalism is a legitimate concept concerning scientific methodology but *only*—and I reemphasize, *only*—if certain limitations on how it is applied in the natural sciences are respected. First, it must respect and adhere to a rational meaning of *natural causes* inferred from *observed effects* rather than *metaphysical implications*. Second, the legitimate domain of methodological naturalism is constrained to the domain of the natural sciences, the material world. Methodological naturalism can only go where science can go. It is perfectly legitimate to acknowledge the metaphysical implications that might result from a scientific observation, regardless of the nature of those implications, teleological or materialistic, but those implications cannot be used to establish whether or not the methodology or the conclusions that resulted from that observation was "scientific" or not. Specifically, methodological naturalism, in considering some aspect of scientific investigation, cannot be accepted as valid unless it explicitly distinguishes between the metaphysical implications of a theory and the methodology employed is the development of that theory.

There is every indication of supernatural causation behind the cosmos, but whether there is or whether there isn't, is irrelevant to the methods and purposes of science.

341

> *The legitimate concern of methodological naturalism within the natural sciences is natural causation, not the ontological properties of the natural world. When science gives metaphysical implications priority over scientific evidence rationally inferred through Bacon's observational, experimental, and inductive methodological tradition, it has abandoned the realm of science for the realm of metaphysics.*

Design in nature points to a *Designer* behind nature just like the appearance of law in nature points to a *Law-Giver* behind nature. Further, we find that law and design are unified in nature around a single organizing principle, the centrality of the place of man in the universe. The natural sciences have revealed that the natural world is teleological through and through. The whole charade of methodological naturalism and its covert mission of eliminating teleological thinking in the natural sciences comes tumbling down in the face of these observations. Methodological naturalism is a sham. It is *stealth naturalism*. Its covert mission is to eliminate teleological thinking from the natural sciences in a world that is teleological through and through.

CHANCE CAUSES (LUCK)

The two great explanatory principles of the natural world are not *law* and *chance* but *law* and *design*, and they are unified in the natural world around the centrality of the place of man in the universe. In this causal system, *chance* is still important, but it is axiomatic that in a rational universe *chance* must forever be confined to the limitations imposed by the physical laws and principles of the system within which it operates. No rational argument or experiment has ever proven otherwise.

While some will note the supposed central role of chance in quantum mechanics, this is only an illusion, an artifact of the positivist epistemology used in the construction of quantum theory. An appeal to quantum uncertainties as a cause is meaningless.

In reality, chance and law or "chance and necessity" as Jacque Monod would say,[3] can be more appropriately described as "luck and necessity." If you win big in Vegas at the slots, it is *luck* that explains the smile on your face and the jingle in your pocket, not *chance*.

NATURAL CAUSES, UNIFICATION, AND THE GOSPEL OF THE COSMOS

My summary here brings together the principle arguments of this treatise concerning the natural sciences. In opposition to the materialism that rules the academy today, I believe these arguments provide both a compelling and perhaps even a conclusive case for the teleological view of the world. Regardless, the thin ice that scientific materialism rests on just cracked a little more. The *Big Splash* is coming.

It is claimed that science can only investigate that which it can observe, and since science cannot observe supernatural causes, it must limit its efforts to the pursuit of only natural causes. However, there are essential elements of nature, which are immaterial and consequently unobservable, such as physical laws, principles, and mechanisms, which exist as abstract conceptions of the mind. Nevertheless, they are *present* and *operative* in the natural world and must be regarded as real parts of the natural world. We know that they are *present* and *operative* because we observe their effects. This meaning of *natural causes* is rooted in actual scientific reasoning applied to the natural world rather than metaphysical criteria. We conclude that:

> *Natural causes are those causes which produce observable effects in the natural world that can be subjected to scientific investigation. Science observes effects and infers causes. The nature of the effect reflects the nature of its cause.*

Some things in nature reflect chance-like properties, some things in nature reflect law-like properties, and other things in nature reflect design-like properties. How can you argue that *law* in nature is a valid scientific concept, but *design* in nature is not when both are inferred from the nature of the distinctive observable effects they produce, whether law-like effects or design-like effects? As for *chance,* given a natural system (such as the natural world), *chance* applies to such systems but only within the range of outcomes permitted by the laws and principles of that system. It is the laws and principles of the system that determine what can and cannot happen, not chance.

The work of a cosmic Law-Giver and Designer behind nature is overwhelming. You cannot make sense of the natural world apart from these two concepts acting together under some common plan and purpose, activities impossible apart from intelligent agency. Just as we bring together and unify our understanding of natural laws and design principles in human contrivances (engineering design), so also we see the same unification of law and design principles in *naturally engineered systems* that far exceed the capability of human engineers.

We thus conclude that whatever scientific status is assigned to *law,* it must also be assigned to *design.* If *law* is "scientific" then *design* is "scientific." If *law* is accepted as a natural cause, so also must *design* be accepted as a natural cause. Or if you insist that *design* is "religion," then you have no choice but to conclude that *law* is also "religion." After all, the idea of a Law-Giver is as old or older than the Old Testament, which, as we are frequently reminded, concerns *religion,* not *science.*

There are profound implications for the meaning of *natural causes* posited in this way, especially when we acknowledge that this understanding is affirmed through our analyses of *naturally engineered systems.* In these naturally engineered systems, we see the effects of *law* and *design* in nature not as two isolated and non-interacting causal systems, but rather, as in systems designed through human engineering, we see them as unified around some common purpose to serve some intended end. The unifying principle of the cosmos that has emerged in the last 100 years as a consequence of scientific discovery in the biological and cosmological sciences is that law and design are unified in nature around the centrality of the place of mankind in the universe. *Mankind is central to the meaning and purpose of the universe.* Contrary to the Copernican principle, it seems that mankind just might be special after all.

There are further practical implications that arise as a consequence of these conclusions.

> *If law and design are unified in nature, then they must also be unified in the natural sciences, and if they are unified in the natural sciences, then they must also be unified in science education in public schools.*

The question of the ages for philosophers and theologians has been "What is the relationship between mankind and the universe?" We now have an answer to that question, one that comes to us not as religious doctrine or philosophical speculations but as logical inference from the natural sciences. The discovery of purpose and intent in the universe by means of the natural sciences is a stunning development in the history of mankind. Mankind is central to the meaning and purpose of the universe. This is the *Gospel of the cosmos*.

PUBLIC EDUCATION

What the last seventy years have taught us about the federal government is that it can no longer be entrusted with matters concerning the education of our children. Both public education and the judiciary have been infiltrated and taken over by those who are committed to cultural transformation and who understand that the key to such transformation is to gain, through public education, access to the minds of our children so that they may be indoctrinated into progressive ideals that are antagonistic to our founding precepts in natural law and Christian moral philosophy. At the heart of that transformation is a materialistic view of man, a progressive political philosophy, a secular moral philosophy, and positive law.

346

Accordingly, following *Everson*, we see the emergence of something like an ad-hoc partnership between public education, the federal judiciary, and the secular scientific establishment to replace God with Darwin. Nothing so effectively undermines the natural law precepts of our founding than the teaching within public education that mankind is an accident of nature.

The key to a liberal education and the preservation of liberty lies in the decentralization of public education to local control in which parents are involved. Of the broad range of topics of concern to the education of our children, it is crucial to the future of the American Republic that our children be schooled in and come to understand and cherish the *American story* in its entirety—the good, the bad, and the ugly. It all needs to come out.

෯   ෯   ෯

Assuming that the present ideological corruption of public education will not suddenly be reformed from within, there are nevertheless general principles that must be respected today by public education concerning science education that respects the methodological norms of science and acknowledges the religious implications inherent to scientific theories concerning the origins of mankind. Public education would demonstrate good faith in its responsibilities by adopting the following principles.

(1) Students should be informed of the fact that there is no war between science and religion and that the "warfare thesis" was a fabrication designed to predispose the minds of educators, politicians, judges, and anyone else of concern against the teleological

view of the world. Contrary to the myth of a war between science and religion, the opposite is true. Modern science emerged out of Christian Europe in the sixteenth and seventeenth centuries where it was the Christian view of the world as a created order that provided the mindset that made science possible. The Christian doctrine of creation presupposes an objective reality that is intelligible to the mind of man. This is not religious doctrine. It is a conclusion of history and natural philosophy.

(2) Students should be informed of the fact that scientific theories about the origins of man have profound religious and political implications for our understanding of ourselves, our place in the universe, our obligations to our fellow man, the kind of political system we choose to live under, how we live our lives, whether we see purpose and meaning to our existence, and our beliefs concerning eternal life. If the state mandates scientific instruction that teaches that mankind is descended from lower forms through unguided material causes alone, then it must fully disclose both the religious and political implications inherent to that theory: man was not created by God, he is a mere accident of nature. Further, natural law, which presupposes a Creator, is a false paradigm for the framing of nations and political systems. To mandate the teaching of a naturalistic scientific theory of man, and then to purposely conceal the religious and political implications of that theory from students, constitutes religious and political indoctrination.

(3) Because of the dominance of Darwinism in the twentieth century and its influence throughout the world, students should be informed as to its moral, cultural, and political consequences, particularly with respect to eugenics and its role in formulating social policies for Hitler's Third Reich and the Holocaust.

(4) If biological evolution is to be taught (and it should), it must be taught objectively without naturalistic or religious bias. This means (a) exposing students to legitimate scientific evidence that both supports and challenges Darwin's claims, (b) informing students that science has no naturalistic account of the origins of life, and without a scientific account of the origin of life, evolution is a theory without a scientific foundation, and (c) exposing students to evidence of design and purpose in the natural world. Evidence of design is ubiquitous throughout the natural world and until the censorship imposed during the last half of the twentieth century, it was a legitimate topic of debate. As shown herein, there is no legitimate scientific or constitutional reason that this information should be withheld from students. To censor scientific evidence of design in nature from science education serves metaphysical and political purposes, not science education.

(5) As an essential aspect of their education in the natural sciences, students should be helped to understand the relationship between cause and effect in the natural world and what constitutes a natural cause. They should be helped to understand that both law and design are natural "causes" and of the same cognitive class.

Like mathematics, they are immaterial conceptions of the mind. Students should be exposed to the rationale that leads to the conclusion that law and design are unified not only in manmade contrivances, but in natural systems as well. Further, students should be informed of the rationale that points to the unification of law and design in nature around a fundamental cosmic principle: the centrality of the place of mankind in the cosmos.

(6) Students should be helped to understand the distinction between the religious implications of a theory or hypothesis of science and the theory or hypothesis itself. Just because evidence of design may be inferred from biological systems in nature, that does not make "design" a religious concept. It may justify the design inference, but that inference itself does not depend on belief in a designer. It is formed as a logical consequence of directly observed scientific information, independent of any preconceived metaphysical notions.

(7) Students should be helped to understand the nature of, and uncertainties that are inherent to, the historical sciences, and the basic distinction that must be made between circumstantial evidence concerning unobserved events that supposedly took place in the remote past and direct evidence based on observations and experiments in the present. There is a fundamental difference in methodology between the science of biological evolution and the science of thermodynamics and physics. To equate them is to deceptively assign to evolutionary biology a scientific status

it does not have and cannot have. It is nonsense to say that the scientific proof of the evolutionary development of mankind is as sound as the scientific proof of the theory of gravity.

THE FUTURE OF THE AMERICAN REPUBLIC

The American Republic was founded on the natural law precepts set down in the Declaration of Independence concerning *the supremacy of the Laws of Nature and of Nature's God, the equality of all men in the eyes of God, the natural Rights given by the Creator to all men, and the place of government ordained by God in protecting the natural Rights of men.* Every aspect of our nation, its Constitution, and its laws and institutions are rooted in these four precepts. In the eyes of the Founders, this framework and the relationships it entailed were simply viewed as part of God's divine plan for men and nations.

These natural law precepts rest on general philosophical considerations regarding the nature of reality and the place of men and nations in that reality—the *first things*. These precepts, while religious in nature, do not constitute a religion in the sense of a particular organized community of believers who hold to a generally agreed-upon set of doctrines and practices. Rather, they constitute a universal declaration of the natural rights of man and establish principles that, while philosophical in nature, can legitimately be called a *natural religion,* a religion which is revealed through nature to all men and understood through natural philosophy completely apart from prophets, divine revelation, and Holy Scripture. This *natural religion* and the universal truths it entails are knowable to all men and subject to none. Some have called it the *religion of America.*[4] This *natural religion* is preexistent to man, it transcends the temporal institutions and religions of man, and it is not subject to the laws of his governments or the opinions of his courts. The Creator is the object of its worship, and its church is the whole realm of creation. The Founders built a nation on this *natural religion*, and we cannot fully understand the intent and vision of the Founders or the ultimate meaning of the Constitution apart from its precepts. Lose this vision, and we have lost the vision of the Founders.

While these *natural truths* revealed in nature are completely consistent with Christianity, they do not depend on Christian doctrines, teachings, or revelation. Nevertheless, it cannot be denied that there is a deep historical connection between Christianity and natural law, and while Christianity is not explicitly written into the Declaration, its ideals about a Creator and the natural rights of man are. This Christian connection to the Declaration and thus to the American founding is not just an incidental or accidental aspect of our history; it is fundamental, giving form, substance, and context to what is meant by religious liberty, moral virtue, and political philosophy.

In the American Republic, the expression of devotion to God within the public sphere flows naturally and freely from the ideals of the Declaration of Independence with its appeal to the laws of nature and of nature's God. God is the Author of the American founding, and as such, it is fitting that we as a nation should publicly offer to the Creator our expressions of gratitude and praise. Within the American Republic, there can be no venue, *public or private,* where, as the occasion may demand, such expressions to God our Creator and Ruler of the Universe are not appropriate, and there can be no *public* venue where, by law, such expressions can be prohibited. The *truth and surety* of these things are as sound as the laws of nature and of nature's God themselves.

Accepting that an individual's relation to God must be a matter of his own volition and that the natural law precepts of the Declaration are applicable to all men, we arrive at a general proposition.

> *Because of our founding in natural law, government in the American Republic must be positive with respect to religion in general and neutral with respect to individual religious sects, but it can never be neutral with respect to God, the Author of our founding.*

The secular doctrine that now rules the judiciary cannot possibly be valid, and neither can the opinions of the courts handed down on the basis of that doctrine be valid. A secular judiciary is a judiciary that by definition has severed its anchor in natural law.

⋖ ⋖ ⋖

That men might write law (as our Founders did) to protect religion is commendable. That jurists might use that same law (as the Supreme Court has done) to banish God and prayer from public schools is seditious. After more than 150 years of First Amendment jurisprudence, we must ask just what it was that convinced the Supreme Court back in mid-twentieth century that nonsectarian prayer to God in public schools was suddenly unconstitutional. Given the First Amendment that says that "Congress can make no law respecting the establishment of religion or prohibiting the free exercise thereof," how does the judiciary prohibit that which the First

Amendment says it is prohibited from prohibiting? This is not a rhetorical question. The Supreme Court actually came up with an answer to (or, a way around) this question and as a consequence, prayer in public schools, a tradition since colonial days, was outlawed.

During the last seventy years (at least) the judiciary has been undergoing a transformation from judicial philosophy rooted in natural law to judicial philosophy rooted in positive law with its irresistible progressive proclivities. The stakes here go far beyond the question of sound judicial principle. The stakes are political power and the transformation of America. The ideals of a progressive society can never be attained under natural law, especially within a culture dominated by a Christian people who understand and are devoted to our founding precepts in natural law. God, natural law, and Christian moral philosophy must be banished, and, as Marxists of the twentieth century found, the place to start is in the classrooms of public education.

Thus, the Court devised a "work-around" to the prohibitions of the Religion Clause. Surprisingly, the solution to this problem was relatively unsophisticated, but if you have raw power at your disposal, you really don't need sophisticated reasoning. You just need a majority.

Relevant case law from *Everson* (and its twisted offspring) can be used to piece together the following scenario for the "work-around". *Accommodation* by the state of prayer in public schools constitutes *establishment* of religion which, under the *Establishment Clause* of the First Amendment, is unconstitutional. Consequently, prayer in public schools is unconstitutional and therefore prohibited. What this bit of judicial *slight-of-hand* reduces to is that "that which under the Free Exercise Clause *cannot* be prohibited, under the Establishment Clause, *must* be prohibited." Or, in its simplest expression, the Establishment Clause was used to countermand the Free Exercise Clause.

Now, it is quite obvious that "accommodation of prayer" by public education does not in fact constitute "establishment of religion" by Congress. Nevertheless, it must be conceded that "accommodation of prayer" by public schools can to some extent have the same effect as "establishment of religion" by Congress, and thus there is some merit to the argument. However, the Founders must have understood the difference between "establishment" and "accommodation" and if it was accommodation they had in mind then they would have said "the congress shall make no law respecting the accommodation of religion . . .". Surely it was never intended that the Supreme Court should be able to use the *Establishment Clause* to countermand the *Free Exercise Clause*, especially when the motivation for doing so was the desire to eradicate God from government. The intent of the Founders was that the *Establishment* and *Free Exercise Clauses* stand alongside one another to defend religious liberty, not *against* one another to destroy religious liberty or to serve the political interest of the judiciary.

One way to resolve this matter is simply to concede that it was never intended that the *Establishment Clause* be used to countermand the *Free Exercise Clause*; therefore, such a scenario is an invalid use of those clauses.

However, there is another factor that can be brought to bear with considerably more weight than merely appealing to good manners and good faith intent. The matter should be, and can be, resolved *in principle* but only when we finally come to understand that the *Establishment* and *Free Exercise Clauses* of the First Amendment pertain to *religion,* not God, and while it could be argued that the difference is merely semantic, that difference is in fact as real as the differences between the Constitution and the Declaration of Independence, or specifically, the First Amendment Religion Clauses and the natural law precepts of our founding. This leads us to the conclusion that while it might be appropriate that the state be neutral with respect to religion, it can never be neutral with respect to God. God is the source of life and liberty, the Author of our founding precepts, and is worthy of our adoration and thanksgiving. Such matters that concern the fundamental relationship between the people and their God cannot be subjugated to the First Amendment and the opinion of the courts.

ఆ ఆ ఆ

If we are to understand the corruption of the federal judiciary that exists today, that corruption must be measured against the standards held in late eighteenth century by the Founders and defined within the providence of natural law and Protestant Evangelical Christianity. If we can properly understand *that* context, then we have a proper understanding of the context within which the First Amendment Religion Clauses were written.

Natural law concerns "knowledge of God taken from nature" and leads to an understanding of the proper temporal relationships between Man, God, and the state. *Religion* concerns "knowledge of God received through revelation" and leads to an eternal relationship between Man and God. Knowing *about* God is philosophy. *Knowing* God is religion. Our founding documents reflect both the philosophical and the religious aspects of God, one in the Declaration and the other in the Religion Clauses of the First Amendment. When it comes to law, they cannot be

356

entangled. Specifically, a precept of the Declaration cannot be entangled with a provision of the Constitution such that that precept is made subject to that provision and thus comes under scrutiny of the Supreme Court.

The objective of *Everson* was the secularization of public education and the indoctrination of future generations so that students will learn at an early age to reject, at the most fundamental level, the founding precepts of the Declaration of Independence and the Christian moral philosophy that have been the heart and soul of the American Republic since its founding. Belief in a Creator based on a conclusion of natural philosophy prepares the way for belief in God, not as a matter of blind faith, but, as Voltaire concluded, as a matter of *reason*. The purpose of public education is, through Darwinism, to sever this connection.

◈ ◈ ◈

Throughout the founding era and the nineteenth century, reference to religion meant Christianity. There was nothing religiously neutral about the spirit of the founding era even for those who did not profess Christianity, and *secular* wasn't even a word in the English language at that time. The religious spirit that energized the founding era was uniquely Christian. A reading of the history of those times does not reflect in any form or to any degree whatsoever the neutrality of *Everson* and *Epperson* or the secularism of *McLean, Edwards,* and *Kitzmiller*—or the lies that undergird *Roe*. In light of the extraordinary sacrifices of those generations who came before in establishing and defending the American Republic, it would seem that the judiciary would strive to honor and protect these treasured principles, honor the sacrifices of our fathers, and above all else, guard the natural rights of man. While the judiciary would argue that religious neutrality and secular purpose within government is the key to guarding our natural rights, in reality what we see is that natural rights today are defined in terms of positivist and progressive ideals, not the laws of God. Ever since *Everson*, it seems that much of the work of both public education and the judiciary has been dedicated to eradicating the natural law precepts of the Declaration from public life and to replace God with Darwin, or at least, use Darwin to minimize the influence of Christian moral philosophy in our culture.

 formula formula formula

There are certain things about our founding that are fixed and not subject to change in spite of how *times* have changed and how our culture has changed and how the opinions of the courts have changed. The religions of men and the cultural norms that we as a nation embrace may change over the years, but the four precepts of *natural law* set down in the Declaration of Independence are ordained by God, and as such, they are universal, eternal, unchanging, and true. They are not incidental to our nation's founding, and they are not optional to our nation's future. They are fundamental. They are the *first things* and our only sure foundation. Any law of Congress or opinion of the courts that is not in accord with these fundamental truths is contrary to our founding.

In the American Republic, we are free to choose the form of religion that we judge to be true and best suited for our lives or no religion at all. What nature teaches us, however, and what the Founders understood, is that in a universe created by the one true God, and designed with mankind

358

as its central purpose (as presupposed in the Declaration of Independence and affirmed in our own times by the natural sciences), reason leads us to conclude (Darwinism and postmodernism notwithstanding) that just as there can only be one set of natural laws that govern the behavior of the natural world, there can only be one ultimate Truth and a natural world that embodies that Truth and which reveals to all men the laws of nature (physical laws), rules for right conduct (a moral philosophy) and the natural rights of man (civil and religious liberties).

The earth in its orbit around the Sun may be regulated by the laws of gravity, but we cannot sit in judgment on those laws. We can only try to understand them, respect them, and perhaps use our knowledge of them to serve some purpose, hopefully good rather than evil. Neither can we sit in judgment of God's laws as they relate to men and nations. We can only respect those laws and order our lives accordingly. In light of scientific discovery today as it pertains to the place of man in the universe, we must conclude that any attempt to secularize government must be based on either a profound ignorance of the precepts of our founding or a profound antagonism toward those precepts. Whether due to ignorance or antagonism (or both), such ideals are invalid in representing the interest of a free people.

∽ ∽ ∽

Everson notwithstanding, it was Christians, not secularists, who came to America seeking religious liberty, and it was the great stirring of Christian faith, not secularism, during the *Great Awakening* in the decades leading up to the Revolutionary War that ignited the fires of liberty in the hearts of the colonists. It was another stirring of Christians, not secularists, known as the *Second Great Awakening* at the end of the eighteenth century that eventually led to the end of slavery in North America, and it was virtue instilled in the people and the institutions of our culture arising from Christian moral philosophy, not secularism, that preserved that liberty to present times.

"In God we trust" is not just a trite saying. It is our foundation. It is what holds us together and the only possible means of withstanding all of the spiritual, cultural and political divisions that try to tear us apart. Our destruction is being engineered by forces that are now at work within our own government and our culture, and if we are to survive the current onslaught, it will only be because we as a nation finally recognize the nature of the conflict, repent, turn away from the false prophets among us and return to God.

It was the pastor-warriors of the *awakened* churches that first *called out* and then *led* the colonists to arms and the establishment of the American Republic. It may be that the only way to save our republic in the twenty-first century is through modern-day pastor-warriors who call us out and lead us, not to the military battlefield, but to a different kind of battle, the battle to preserve the *first things* of the American Republic and our cultural foundation in Christian virtues.

The good news is that the means for our salvation is already in place. It was put there at our founding at great cost, and over the years, it has been defended at great cost. For now, it is simply a matter of electing an overwhelming majority of representatives in our state and federal governments who understand and are committed to the *first things*—just like in 1776. But to *elect* such men and women, we must first *become* such men and women, and out of a community of such men and women, God will call out some to serve.

For the last 100 years, we have been driven into the wilderness by materialistic dogma and positivist epistemology. We need to get classical physics back on the rails (back to realism), we need to get biology back on the rails (back to design), we need to get the judiciary back on the rails (back to natural law), we need to get the American Republic back on the rails (back to the *first things*), and we need to get mankind back on the rails (back to the Creator).

∾ ∾ ∾

Within the traditional meaning of religion in America concerning our worship and service to God, how does one distinguish between a government that operates on secular principles for the sake of religious liberty, and a government that operates on secular principles for the sake of secularism itself? The first seeks *freedom of religion* and publicly honors God as Creator and Author of the natural rights of man and is faithful to its sacred mission of protecting those rights. Under this view, a central goal of public education should be to deeply embed in the hearts and minds of every student the four precepts of natural law—*the supremacy of the Laws of Nature and of Nature's God, the equality of all men in the eyes of God, the natural Rights given by the Creator to all men and the place of government ordained by God in protecting the natural Rights of men.*

The second, however, seeks *freedom from religion* and fulfills that objective by eradicating all traces of God from public life through the establishment within government of a radical secularism. Its principle

361

means for indoctrination is public education where natural law and its four precepts are regarded as no longer being relevant to our modern times or to an enlightened education.

Once freedom *from* religion is securely established within government and our public institutions, it will then be safe to say that this is the turning point, signifying the beginning of the end of the Republic, the birth of Comte's *age of positivism* and the dawning of his *religion of humanity*. Finally, the long awaited utopian age will be realized.

But given the depth of love of God and country that still burns deep within the American Republic, I wouldn't bet on it.

NOTES

PREFACE

1. David Berlinski, "The Deniable Darwin," *Commentary Magazine*, June 1, 1996.

2. David Berlinski and Critics, "Denying Darwin," *Commentary Magazine*, Sept 1, 1996.

3. Metaphysics - A division of philosophy that is concerned with the fundamental nature of reality and being.

4. Among the interpretations of reality that came out of the Enlightenment period, it was Scottish Common Sense Realism that most directly impacted the thinking of the Founders of the American Republic, in contrast for instance, with Kant's views that reality could not be known with certainty because we had no way of knowing if our sensory perceptions actually correlated to reality. Kant was wrong, and his ideas would come to haunt us in the twentieth century in their support of the positivist epistemology.

5. C. S. Lewis' books *Mere Christianity* and *The Problem of Pain* are a good place to start in gaining a sound understanding of Christianity's deeper truths.

6. From "High Flight," a poem written by John Gillespie Magee, Jr. an American fighter pilot serving in England with the Royal Canadian Air Force in the Second World War, killed at the age of nineteen, while flying a British Spitfire.

7. John Lennox, *God's Undertaker: Has Science Buried God?* A Lion Book, 2007.

8. David Berlinski, *The Devil's Delusion,* Crown Forum, 2008.

9. Stephen C. Meyer, *Signature in the Cell,* Harper One, 2009.

10. Stephen C. Meyer, *Darwin's Doubt,* Harper One, 2013.

11. John H. Calvert, *Kitzmiller's Error: Defining "Religion" Exclusively Rather Than Inclusively,* Liberty University Law Review, Volume 3, Number 2, Spring 2009. Calvert published a follow-on article in the Spring 2018 issue of Liberty University Law Review further expounding on the problem of failure of the courts to employ an *inclusive* meaning of religion.

12. Douglas Axe, *Undeniable,* Harper One, 2016.

CHAPTER 1   INTRODUCTION

1. Charles Darwin, *On the Origin of Species,* 1959.

2. Lawrence M. Principe, *History of Science,* "The Great Courses," The Teaching Company [Audio CD], Chantilly, VA, 2002.

3. Ontology (semi-formally): having to do with existence, the essential nature of a thing, a thing's essential properties that make it what it is.

4. Richard Lewontin, "Billions and Billions of Demons" in *The New York Review of Books,* January 9, 1997, Volume 44, Number 1.

5. Alvin Plantinga, *Where the Conflict Really Lies,* Oxford, 2011.

6. Methodological naturalism is a recurring theme throughout this work and was the subject that inspired me to begin writing in the first place. As it turned

out, it was rather easily dealt with and while this is important to practical matters related to the philosophy of science, science education, and First Amendment law, these concerns eventually get swept away as incidental to the larger picture.

7. Michael Denton, *Evolution: A Theory in Crisis,* Adler&Adler, 1986, 358.

8. Teleology - the explanation of phenomena in terms of the purpose they serve rather than of the cause by which they arise.

9. Thomas Woodward, *Doubts about Darwin,* Baker Books, 2003, 31.

CHAPTER 2 METHODOLOGICAL NATURALISM, CONSTITUTIONAL LAW, AND SCIENCE EDUCATION IN PUBLIC SCHOOLS

1. Alvin Platinga, "Methodological Naturalism?" *Origins and Design*, Vol. 18, No. 1 & 2, at www.arn.com. Also at https://www.calvin.edu/ . . . /plantinga_alvin/methodological_naturalism_part_1.pdf and https://www.calvin.edu/ . . . /plantinga_alvin/methodological_naturalism_part_2.pdf. This is a lengthy analysis in two parts by a notable and universally respected Norte Dame philosopher of science, giving a conservative Christian view.

2. Barbara Forrest, "Methodological Naturalism and Philosophical Naturalism: Clarifying the Connection," *Philo* 3 (2):7–29 (2000). Forrest is professor of the philosophy of science at Southeastern Louisiana State University, presenting an extreme naturalistic view. Forrest is a major leader in the American Humanists movement and is often called in as an expert witness in court cases concerning the teaching of evolution in public schools.

3. John Rennie, "15 Answers to Creationist Nonsense," *Scientific American,* July 2002.

4.   Thomas Nagel, *Mind and Cosmos*, Oxford University Press, 2012.

5.   Iris Fry, "Is science metaphysically neutral?," *Studies in History and Philosophy of Biological and Biomedical Sciences*, Department of Humanities and Arts, Technion-Israel Institute of Technology 43 (2012) 665–673. Fry also authored a book on this subject, *The Origins of Life on Earth*, Rutgers University Press, 2000. Nice to run across a naturalists who is willing to tell the truth about the role of methodological naturalism in the natural sciences.

6.   Calvert, op. cit.

CHAPTER 3   THE NATURAL SCIENCES

1.   Nancy R. Pearcey and Charles B. Thaxton, *The Soul of Science*, Crossway Books, 1994. There is a lot of meat in this little book, and Nancy does a superb job of addressing the main aspects of the emergence and development of science within the biblical view of the world.

2.   J. P. Moreland, *Christianity and the Nature of Science*, Baker Books, 1989.

3.   Lawrence M. Principe, *History of Science*, The Great Courses, The Teaching Company [Audio CD], Chantilly, VA, 2002.

4.   James Hannam, *The Genesis of Science, How the Christian Middle Ages Launched the Scientific Revolution*, Regnery, 2011.

5.   The history and philosophy of science as well as scientific methodology are all interrelated. For anyone wishing to get a decent introduction to this topic, the four preceding citations should serve that purpose. This is a tremendously fascinating and important subject, providing insight into our understanding of the relationship between science and Christianity as well as the role of methodological naturalism in the natural sciences.

6. The patristic writings consisted of the writings of the church fathers during the first 400 years after Christ, an exceedingly important period during which they sought to establish the doctrines and practices of the church within a Greco-Roman culture.

7. Nagel, op. cit.

8. R. P. Feynman, R. B. Leighton, M. Sands, *The Feynman Lectures on Physics,* Vol. 1, Addison-Wesley, Reading, MA, 1963.

9. Ernst Mayr, "Darwin's Influence on Modern Thought," *Scientific American,* July, 2000.

10. See http://www.copeinc.org/science-readings.html for information regarding Next Generation Science Standards and how they effectively constitute establishment of a naturalistic religion in public education.

11. For instance, see Jonathan Wells' two books, *Icons of Evolution, Science or Myth?,* Regnery Publishing, 2000 and *Zombie Science, More Icons of Evolution,* Discovery Institute Press, 2017.

12. John W. Oller Jr. and John L. Omdahl, "Origin of Language Capacity: in Whose Image?," in *The Creation Hypothesis,* Edited by J. P. Moreland, Intervarsity Press, 1994, 241.

13. Cornelius G. Hunter, *Science's Blind Spot,* Brazos Press, 2007.

14. C. S. Lewis, Mere Christianity, Harper One, 1952, 39.

CHAPTER 4   THE "WAR" BETWEEN SCIENCE AND RELIGION

1. Lawrence M. Principe, *Science and Religion,* The Great Courses, The Teaching Company, [Audio CD] Chantilly, VA.

2. Giorgio de Santillana, *The Crime of Galileo,* University of Chicago Press, 1955. The Galileo affair is a popular subject and the object of the work of

numerous authors. This book by Massachusetts Institute of Technology professor of history and the philosophy of science Giorgio de Santillana (deceased in 1974) is regarded as the benchmark for modern scholarship on this subject.

3. John William Draper, *History of the Conflict between Religion and Science,* publisher unknown, 1874.

4. Andrew Dixon White, *A History of the Warfare of Science with Theology in Christendom,* Volumes I and II, Prometheus Books, 1993, originally published by Appleton, 1896.

5. Rodney Stark, *The Victory of Reason,* Random House, 2005, 35–36, 12–16.

6. Edward J. Larson, *Summer for the God*, Harvard University Press, 1997.

7. God & Nature, *Historical Essays on the Encounter between Christianity and Science*, Edited by David C. Lindberg and Ronald L. Numbers, University of California Press, 1987. A benchmark in dispelling the warfare mythology.

8. *Science and Religion: A Historical Introduction,* edited by Gary B. Ferngren, The Johns Hopkins University Press, 2002.

9. Ronald L. Numbers, "Science Without God: Natural Law and Christian Beliefs," in *When Science & Christianity Meet,* Edited by David C. Lindberg and Ronald L. Numbers, The University of Chicago Press, 2003, 265–285.

10. Views of Nobel laureates in science, Nobel authors, and Nobel Peace laureates can be found at http://nobelist.net.

11. Michael Ruse, *The Creation Struggle,* Harvard University Press, 2005.

12. Alvin Plantinga, *Where the Conflict Really Lies,* Oxford, 2011.

CHAPTER 5   SCIENCE AND NATURALISM

1. Larson, E. J. & Witham, L. *Nature* 386, 1997, 435–436.

2. Douglas Futuyma, *Evolutionary Biology,* 3rd Ed., Sinauer, 1998.

3. John Searle, *The Rediscovery of the Mind,* MIT, 1992, 3–4.

4. Franklin Harold, *The Way of the Cell, Oxford,* 2003, p. 205

5. Will Provine, quoted at http://bevets.com/equotesp5.htm.

6. Daniel Dennett, *Darwin's Dangerous Idea,* Simon & Schuster, 1995.

7. Jacques Monod, *Chance and Necessity,* Vintage Books, 1972.

8. Benjamin Wiker, *Moral Darwinism,* InterVarsity Press, 2002, 15–50.

9. Richard Bozarth, "The Meaning of Evolution" in the *American Atheist,* February 1978.

10. John J. Dunphy, "A Religion For a New Age," in *The Humanist* magazine, Jan-Feb 1983, Vol. 32, No. 1.

11. See http://thesciencenetwork.org. Click on [meetings] then [beyond belief]. Scroll down to "Science, Reason, Religion & Survival" and view Session 1. You will see that the subtitle "beyond belief" is appropriate.

CHAPTER 6   NATURAL CAUSES

1. In this chapter my development of what is meant by "natural causes" is critical to conclusions made later concerning methodological naturalism as well as the concepts of *law* and *design*. These developments are constructed fully within a general understanding of the history, philosophy, and nature of the natural sciences as developed and discussed in chapters 3, 4, and 5. I rely on few references here in that the particular arguments I present and conclusions I reach are as far as I can tell, my own. Reason leads to the conclusion that the meaning of *natural causes* developed here should be broadly adopted by philosophers of science and incorporated in ongoing discussions regarding

the relationship between science and Christianity, the teaching of the natural sciences in public schools, and First Amendment jurisprudence.

2. Uri Alon, *An Introduction to Systems Biology: Design Principles of Biological Circuits,* 1st Ed., Chapman and Hall/CRC, 2006.

CHAPTER 7 THE INTELLIGENT DESIGN MOVEMENT

1. Sir Julian Huxley, "Evolution After Darwin," ed. Sol Tax, in *The Evolution of Life*, vol.3, The University of Chicago Press, 1960.

2. Stanley L. Miller, "Production of Amino Acids Under Possible Primitive Earth Conditions," *Science. 117 (3046): 1953,* 528–9.

3. Thomas Woodward, *Doubts About Darwin,* Baker Books, 2003, 37–45. Woodward provides a concise summary of the history of the problems that emerged in evolutionary theory during the 1960s, '70s and '80s.

4. *Ibid.*

5. *Ibid.*

6. *Newsweek,* November 3, 1980.

7. Roger Lewin, "Evolutionary Theory Under Fire," *Science,* November 21, 980.

8. Francis Hitching, *The Neck of the Giraffe,* Signet, 1983.

9. G. R. Taylor, *Great Evolution Mystery,* Harper & Row, 1983.

10. Woodward, op. cit.

11. Woodward, op. cit.

12. Stephen J. Gould, "Evolution's erratic pace," *Natural History,* vol. 86, 1977.

13. Niles Eldredge and Ian Tattersall, *The Myths of Human Evolution,* Columbia University Press, 1984.

14. Charles B. Thaxton, Walter L. Bradley, and Roger L. Olsen, *The Mystery of Life's Origins,* Lewis and Stanley, 1992.

15. Robert Shapiro, *Origins: A skeptic's guide to the creation of life on Earth,* Summit Books, 1986.

16. Stephen J. Gould, *Wonderful Life,* 1989, p. 59.

17. Douglas Futuyma, *Evolutionary Biology,* Third Ed., 1998, p. 710.

18. Robert L. Carroll, *Patterns and Processes of Vertebrate Evolution,* Cambridge Paleobiology Series, Cambridge University Press, 1998, pp. 1–18.

19. Campbell and Reese, *Biology,* 7th edition, Pearson Cummings, 2005.

20. Michael Denton, op. cit., 1986.

21. Phillip E. Johnson, *Darwin on Trial,* Intervarsity Press, Second Edition, 1993.

22. Michael L. Behe, *Darwin's Black Box,* Free Press, 1996.

23. William A. Dembski, *Intelligent Design,* IVP Academic, 2002.

24. Richard Dawkins, *The Blind Watchmaker,* Norton, 1986.

25. Douglas J. Futuyma, *Evolutionary Biology,* Third Edition, Sinauer Associates, 1988, 5.

26. Daniel C. Dennett, "Show Me the Science," Op Ed Contributor, *New York Times,* Aug 28, 2005.

27. Francis Crick, *What Mad Pursuit,* Basic Books, 1990.

28. Francisco J. Ayala, "Darwin's Revolution: From Natural Theology to Natural Selection," in *Creative Evolution,* edited by John H. Campbell and J. William Schopf, Jones & Bartlett Publishers, 1994.

29. Ernst Mayr, op. cit.

30. Cambridge University paleontologists Simon Conway Morris in his book *Life's Solutions,* Cambridge University Press, 2003, makes reference to computer simulations demonstrating the "eerie perfection" of the biological code.

31. James A. Shapiro, *Evolution: a view from the 21st century,* FT Press, 2011.

32. Thomas Nagel, op. cit.

33. For instance, see recent Discovery Institute online article https://evolutionnews. org/2017/08/evolutionary-theorist-concedes-evolution-largely-avoids-big- gest-questions-of-biological-origins/ reporting the results of a meeting of the Royal Society in November 2016 entitled "New Trends in Evolutionary Biology." It is now broadly admitted that challenges to the classical model . . . are widespread . . . and none are unscientific." For attendees and abstracts, see https://royalsociety.org/science-events-and-lectures/2016/11/evolutionary-bi- ology/. Note that Denis Nobel was in attendance.

34. Denis Nobel, *Dance to the Tune of Life*, Cambridge University Press, 2017.

35. J. Scott Turner, *Purpose and Desire,* Harper One, 2017. For years I have had the idea that there must be some sort of resident intelligence at work in living things in that it seems to me that there can be no other way to account for the extraordinary integration and cooperation that takes place in real time, both vertically and horizontally, throughout the organizational structure of living organisms. Somehow, someway every cell and every system and subsystem in a living organism "strives" toward a common goal: the preservation of the life of the organism itself. Turner forces us to rethink our understanding of what life really is.

36. Center for Science and Culture, https://www.discovery.org/id/.

37. Access Research Network, www.arn.org.

CHAPTER 8   THE DARK SIDE OF DARWINISM

1. Available   at   http://galton.org/books/hereditary-genius/text/pdf/galton- 1869-genius-v3.pdf.

2. Charles Darwin, *The Descent of Man, and Selection in Relation to Sex,* Princeton University Press (1981).

3. A good bit of what I have written here in this section (composed of the immediate six paragraphs) was extracted from a book review of Edwin Black's 2003 book about eugenics, *War Against the Weak*. I am embarrassed to have to confess that after months and months of searching through boxes and boxes of notes, magazines, and other reference materials (I never throw away anything) I cannot find that reference. So, to the unknown author of that review, I must say that your review was so well written, so concise and powerful, that except for some condensation, I used it pretty much as you wrote it. I thank you whoever you may be.

4. Richard Weikart, *From Darwin to Hitler,* Palgrave Macmillan, 2004.

5. Edwin Black, *War Against the Weak,* Four Walls Eight Windows, 2003. See http://www.waragainsttheweak.com.

6. See Benjamin Wiker, *10 Books that Screwed up the World,* Regnery, 2008. See chapter 7 where he reviews Darwin's second work, *The Descent of Man.*

7. Dennis Sewell, *The Political Gene,* Picador, 2009. Sewell is a writer and broadcaster and a contributing editor to the *Spectator.*

8. *Ibid.* 233–234.

CHAPTER 9   DARWIN, DESIGN, AND THE PHILOSOPHERS

1. Thomas Jefferson, letter to John Adams, from Monticello, April 11, 1823; Lester J. Cappon, ed., *The Adams-Jefferson Letters: The Complete Correspondence*, Chapel Hill and London: University of North Carolina Press, 1987, pp. 591–594.

2. Thomas Paine, *The Study of God,* A discourse delivered to the Society of Theophilanthropists at Paris, Stedman and Hutchinson, comps., A Library of

American Literature: An Anthology in Eleven Volumes, Vol. III: *Literature of the Revolutionary Period*, 1765–1787.

3. Voltaire, Letter to Frederick the Great, 1770.

4. Voltaire, Philosophical Dictionary, Vol. III., Faith, Section I., 1824, 155.

5. David Hume, *The Natural History of Religion,* INTRODUCTION, at http://oll.libertyfund.org/titles/hume-the-natural-history-of-religion.

6. Ibid. Section 15, GENERAL CORROLARY.

7. https://en.wikipedia.org/wiki/Dialogues_Concerning_Natural_Religion.

8. https://plato.stanford.edu/entries/kant-religion/.

9. S. Gaukroger, *Kant and the nature of matter: Mechanics, chemistry, and the life sciences, Stud Hist Philos Sci.* 2016 Aug; 58:108–14. doi: 10.1016/j.shpsa.2016.03.007. Epub 2016 Apr 16.

10. William Paley, *Natural Theology,* DeWard Publishing, 2010.

11. Elliot Sober, *Philosophy of Biology,* Westview Press, 1993.

12. Bertrand Russell, *History of Western Philosophy,* [1945] Simon & Schuster/Touchstone, 1967.

13. Hunter, op. cit.

14. Joel S. Swartz, "Darwin, Wallace, and Huxley, and Vestiges of the Natural History of Creation," *Journal of the History of Biology*, March 1990, Volume 23, Issue 1, pp 127–153.

CHAPTER 10   COSMOLOGICAL DESIGN

1. Denton, M. J. (2013) "The Place of life and man in nature: Defending the anthropocentric thesis." *BIO-Complexity* 2013 (1):1–15. Doi:10'5048/BIO-C:2013.1.

2. Thomas H. Huxley, *Evidence as to Man's place in nature*, Williams & Norgate, 1863.

3. Freeman Dyson, "Energy in the Universe," *Scientific American*, September 1971, pp. 51–59.

4. Alfred Russell Wallace, *Man's Place in the Universe*, Chapman and Hall, 1904.

5. Lawrence J. Henderson, *The Fitness of the Environment*, The Maximillian Company, 1917.

6. Peter D. Ward and Donald Brownlee, *Rare Earth*, Copernicus, 2000.

7. Michael J. Denton, *Nature's Destiny*, Free Press, 1998.

8. Robert Jastrow, *God and the Astronomers*, Norton & Company, 1992.

9. George Greenstein, *Symbiotic Universe: Life and Mind in the Cosmos*, Morrow, 1st Edition, 1988.

10. Quoted by Robert Jastrow in *God and the Astronomers*.

11. Margenau, H. and R.A. Varghese, ed. 1992. *Cosmos, Bios, and Theos,*. La Salle, IL, Open Court, p. 83.

12. Edward R. Harrison, *Masks of the Universe*, Cambridge University Press, 2011.

13. Fred Hoyle, *The Universe: Past and Present Reflections*, Annual Reviews of Astronomy and Astrophysics, 20 (1982), 16. Published in *Engineering & Science*, November 1981.

14. https://en.wikipedia.org/wiki/Infinite_monkey_theorem.

15. George Wald, "The Origin of Life," *Scientific American*, August, 1954.

16. These comments are not word-for-word quotes from Wald in his *Scientific American* article. Rather they are a paraphrase of his description of a dilemma. When it comes to the question of the origins of life, he is caught between having to believe in God or believe in the impossible, spontaneous generation.

But the way out of the dilemma, he says is *time* where given enough time, "the possible becomes probable and the probable virtually certain."

17. Denton, op. cit., 1986.

18. See http://2012daily.com/?q=node/13 for a brief history of the transformation of George Wald.

19. Paul Davies, *The Goldilocks Enigma,* Penguin, 2006.

20. Robert Jastrow, op. cit.

21. Thomas Huxley, letter to Charles Kingsley, September 23, 1860. Kingsley was a prominent clergyman of the Church of England who had written Huxley and his wife on the occasion of the death of their young son. The quote is taken from Huxley's letter in response to Kingsley. Given the circumstances, one cannot help being moved by the respect and compassion shared between these two men in spite of their ideological differences.

CHAPTER 11 UNIFICATION

1. I cannot overstate the influence of Michael Denton on my understanding of unification. It was his recognition in *Nature's Destiny* concerning the centrality of the place of man in the universe that inspired this chapter on unification and the title to my book.

2. The general notion of the significance of location relative to the center of the universe was the question at stake concerning the so-called "Copernican Revolution." However, it turns out that another aspect of our particular location is that we are so positioned in our galaxy and our galaxy is so positioned within the local group that we have an unobstructed view of the universe. These remarkable facts are described by Gonzales and Richards in their book *The Privileged Planet* where they show it was only due to our physical

location in the cosmos that we were able to make the observations and discoveries necessary to the development of the natural sciences, a profound higher-level anthropic coincidence, one with a distinct purpose.

3. Denton, op. cit., 1998.

4. A. J. Gurevich: *Categories of Medieval Culture,* Routledge and Keegan Paul, 1985.

CHAPTER 12   PHYSICS DERAILED

1. To be complete classical physics also includes the thermodynamics of Gibbs, the elasticity and hydrodynamics of La Grange and Hamilton, and the molecular statistics of Maxwell and Boltzmann.

2. I use "our" in a general sense. I don't mean the scientific community as if I was speaking for science itself. I mean *humanity* collectively because all humanity is affected. We really are all in this together.

3. Karl Popper, *The Logic of Scientific Discovery,* Routledge, 1992, 14. Popper's analysis of the assault of positivism on the inductive methodological tradition of classical physics is scattered throughout the first 100 pages of his 500-plus page work. It is summarized very well however, on p. 14 where he says "*the verdict of the positivist dogma of meaning is that both [induction and positivism] are systems of meaningless pseudo-statements. Thus instead of eradicating metaphysics from the empirical sciences, positivism leads to the invasion of metaphysics into the scientific realm.*"

4. Thomas S. Kuhn, *The Structure of Scientific Revolution,* University of Chicago Press; 2nd edition, 1970.

5. Patrick Johnson Mendie & Edwin Ejesi, "Logical Positivist versus Thomas Kuhn," *ResearchGate,* September 2014, at https://www.researchgate.net/publication/ 303784859_LOGICAL_POSITIVIST_VERSUS_ THOMAS_KUHN.

6. Emeritus Professor of Philosophy at Western Washington University Richard Purtell, in the first chapter of his book *Reasons to Believe, Why Faith Makes Sense,* Ignatius Press, 2009, gives a very understandable account of the tautological nature of positivism and its impossibility as a rational theory of knowledge. For the serious student who wants a solid scholarly treatment of why *faith makes sense,* I highly recommend this book. If you are a student in the university, regardless of your area of study, it will serve you well.

7. Lee Smolin, *The Trouble with Physics,* First Mariner Books, 2007.

8. J. R. Carnes, *Axiomatics and Dogmatics,* Oxford University Press, 1982. This example was used by Carnes in his first chapter on "World and Theory," an insightful discourse of the role of theory in science, 1–17.

9. Abraham Pais, *Subtle is the Lord: The Science and Life of Albert Einstein,* Oxford University Press, 2005.

10. Einstein's actual language in his letter to Born on Dec 4, 1926 was "The theory yields a lot, but it hardly brings us any closer to the secret of the Old One. In any case I am convinced that *He* does not throw dice."

11. https://www.sciencenews.org/blog/context/ why-quantum-mechanics-might-need-overhaul.

12. For instance, in some cases in the quantum world, it appears that effect precedes cause.

13. William Cowper and Reverend T. S. Grimshawe, *The Life and Works of William Cowper,* Wentworth Press, 2016.

14. You may wonder what makes me think that I, a humble engineer and confessed nonacademic, can speak with such confidence in my critique of quantum mechanics. To me, this is overwhelmingly obvious for the reasons I have stated. But I have not told the whole story. If there is an opportunity and the time is right to write a second edition to this book, I will finish the story, and you will find out that the process of getting atomic physics back on the rails is already well underway.

CHAPTER 13   THE AMERICAN REPUBLIC

1. Denton, 1986, op. cit., 358.

2. I have relied primarily on four sources regarding this review of the founding of the American Republic. I found these works to be quite consistent and complementary. They are: (a) Steven Waldman, *Founding Faith,* Random House, 2008; (b) Daniel N. Robinson, American Ideals: Founding a "Republic of Virtue," The Teaching Company, Audio CDs, 2004; (c) Allen C. Guelzo, The American Mind, The Teaching Company, Audio CD, 2005; and (d) Paul Johnson, *A History of the American People,* HarperPerennial; 1997.

3. Martin Luther King, Jr., *Letter from a Birmingham Jail,* April 16, 1963, online at Martin Luther King Center, http://www.thekingcenter.org/archive/document/letter-birmingham-city-jail-0.

4. These four precepts concerning the American founding will be repeated verbatim several times throughout the remaining chapters of this book. In recognition of the profound significance of these precepts and the way they

connect us as a nation to the Creator, I follow the convention for capitalization set down by Jefferson in the Declaration of Independence.

5. Daniel N. Robinson, op. cit.

6. W. Cleon Skousen, *The Five Thousand Year Leap*, American Documents, 2009, 41–46.

7. Kenneth Scott Latourette, *A History of Christianity,* Harper, 1953, 957–964.

8. Paul Johnson, op. cit., 109–117.

9. Benjamin Rush on education, 1798, at http://ringthebellsoffreedom.com/Quotes/brushcontent.htm.

10. See *National Archives,* Founders Online at https://founders.archives.gov/documents/Madison/01–08–02–0163.

11. Matthew 22:21, KJV.

12. In the founding era, there was no word *secular*, and in that day they had no other word that reflects the meaning that is commonly accepted today. Its meaning is associated with the stance assumed by government and government activities which are constrained by the First Amendment Religion Clause. Secular implies that such activities are neutral with respect to religion, neither *for* nor *against*. As explained by Benjamin Wiker, it was actually conceived as a "sugar-coated" word to gently convey the ideals of an atheistic philosophy to Christians without giving away its true nature. For details on the origins of "secular" see Benjamin Wiker, *Worshiping the State*, Regnery, 2013, 233.

13. Kenneth Scott Latourette, op. cit., 963.

14. Allen C. Guetzo, op. cit.

15. It is no small undertaking to sort out the effects that the ideas and movements that came out of the European Enlightenment in the eighteenth and

nineteenth centuries had on modern times. Nancy Pearcey, in her book *Total Truth* (Crossway, 2004) approaches this question from the perspective of how those ideas influenced and challenged the Christian view of the world. Benjamin Wiker, in his book *Worshipping the State* (Regnery, 2013) looks at the political impact on the United States here in late twentieth century and the early decades of the twenty-first century.

16. Benjamin Wiker, *10 Books that Screwed up the World,* Regnery, 2008.

17. Commentary here on progressivism in America is based largely on the brief essay by Ronald J. Pestritto, Hillsdale College, *Progressivism and America's Tradition of Natural Law and Natural Rights* at http://www.ninrac.org/critics/american-progressivism. This is an excellent paper on an extraordinarily important subject with links to in-depth essays on this subject. You just cannot understand American politics today without understanding progressivism in America.

18. Humanist Manifesto III at www.AmericanHumanist.org.

19. Benjamin Wiker, op. cit., 233.

20. It is important to understand these three principles, where they came from and their function in forming the moral philosophy of today's secular society. See Pearcey and Wiker at 15 above.

21. Daniel Dreisbach, "The Mythical 'Wall of Separation': How a Misused Metaphor Changed Church-State Law, Policy, and Discourse," The Heritage Foundation at https://www.heritage.org/political-process/report/the-mythical-wall-separation-how-misused-metaphor-changed-church-state-law, June 23, 2006. Dreisbach provides informative historical context and insight contributing to our understanding of these early years of the Republic during Jefferson's presidency. He elaborates on many of the problems caused by

Jefferson's metaphor and its devastating impact on First Amendment jurisprudence in modern times.

22. *Everson v. Board of Education of the Township of Ewing,* 330 U.S. 1 (1947).

23. Benjamin Wiker, *Worshiping the State,* Regnery, 2013. In Chapter 15, pp.
289–300, Wiker makes some compelling remarks about the dire circumstances that resulted from the 1947 Supreme Court case *Everson v. Board of
Education,* the effect of which was to establish the religion of liberal secularism within government. As you will see, I come to the same conclusion but
with a twist. Wiker also offers a bold program in Chapter 16, pp. 303–330, of
disestablishment of that naturalistic religion within government.

24. Skousen, op. cit., 248–251. Skousen provides a brief review of several other
relevant cases not addressed here.

25. *Engel v. Vitale,* 370 U.S. 421 (1962).

26. *School Dist. of Abington Tp. v. Schempp,* 374 U.S. 203 (1963).

27. Wallace v. Jaffree, 472 U.S. 38 (1985).

28. *Lee v. Weisman,* 505 U.S. 577 (1992).

29. *Roe v. Wade,* 410 U.S. 113 (1973).

CHAPTER 14   THE FIRST AMENDMENT AND PUBLIC SCHOOLS

1. *Legal Positivism,* Stanford Encyclopedia of Philosophy, at https://plato.
stanford.edu/entries/legal-positivism/.

2. *Torcaso v. Watkins,* 367 U.S. 488, 495 n. 11, (1961).

3. *Elk Grove Unified School District et al. v. Newdow et al.* (June 14, 2004).

4. Calvert, op. cit.

5. *Epperson v. Arkansas,* 393 U.S. 97 (1968).

6. *Rosenberger v. Rector and Visitors of Univ. of Va.,* 515 U.S. 819, 846 (1995).

7. *Allegheny County v. Greater Pittsburgh ACLU*, 492 U.S. 573, 592 (1989).

8. *Gillette v. United States*, 401 U.S. 437, 449–50 (1971).

9. *Edwards v. Aguillard*, 482 U.S, 578 (1987), quoting *Harris v. McRae*, 448 U.S. 297, 319 (1980) quoting *McGowan v. Maryland*, 366 U.S. 420, 442 (1961).

10. *Malnak v. Yogi*, 440 F.Supp. 1284 (1977).

11. *Lemon v. Kurtzman*, 403 U.S. 602 (1971).

12. Ibid.

13. *School Dist. Of Abington Tp. v. Schempp* 374 U.S. 203 (1963).

14. *Crockett v. Sorenson*, 568 F. Supp. 1422 (1983).

15. *Epperson,* op. cit.

16. *Edwards v. Aguillard*, 482 U.S, 578 (1987).

17. Available   at   https://archive.org/details/362091-louisianas-balanced-treatment-for-creation, published May 29, 2012.

18. *McLean v. Arkansas Board of Education*, 529 F. Supp. 1255 (1982).

19. Justice Black, in his concurring opinion in *Epperson,* raised legitimate doubts as to whether or not this case should have even been heard by the Court. It seemed that the supposed controversy was somewhat fabricated. There was little to suggest that Susan Epperson would actually be disciplined if, as prescribed by the state, she taught the subject matter in the new textbook concerning evolution. This case was to a large extent nothing more than an exercise in judicial theory, an exercise, however, that was useful to the evolutionists, thus paving the way for more substantive cases. The question of the teaching of evolution versus creation was now on the radar screen of the Supreme Court.

20. *Wallace v. Jaffree*, 472 U.S. 38, 105 S. Ct. 2479 (1985).

21. *But is it Science? The Philosophical Questions in the Creation/Evolution Controversy,* Edited by Michael Ruse, Prometheus, 1996.

22. Ibid., 337–350.

23. Francis J. Beckwith, *Law, Darwinism and Public Education,* Rowman and Littlefield, (2003), 49–89.

24. Michael Ruse, "How evolution became a religion: creationists correct?," *National Post,* May 13, 2000.

25. *Kitzmiller v. Dover Area Sch. Dist.,* 400 F. Supp. 2d 707 (M.D. Pa. 2005).

26. David K. DeWolf, et al., *Traipsing into Evolution,* Discovery Institute Press, 2006.

27. But is it Science? Op. cit., 355.

28. *Zorach v. Clauson,* 343 U.S. 306, 72 S. Ct. 679,96 L. Ed. 954 (1952).

29. *McGowan v. Maryland,* 366 U.S. 420 (1961). It should be noted however, Justice Douglas, in spite of his clear understanding of and commitment to the founding precepts held by the Founders as so elaborately expressed in his dissenting opinion in McGowan, in the years to follow his views, rapidly descended into a radical liberalism that offended even the most liberal of the other justices on the Court.

30. Wallace, op. cit.

31. Calvert, op. cit.

32. *Roe v. Wade,* 410 U.S. 113 (1973).

33. Joseph W. Dellapenna, *Dispelling the Myths of Abortion History,* Carolina Academic Press, (2006).

CHAPTER 15   SUMMARY

1. Alexus de Tocqueville, *Democracy in America,* Alfred A. Knopf; Reprint edition (May 10, 1994).

2. See Note 16, Chapter 10.

3.   Monod, op. cit.

4.   Skousen, op. cit., 65, Skousen's commentary on Alexis de Tocqueville, *Democracy in America,* Vintage Books (1945).

CPSIA information can be obtained
at www.ICGtesting.com
Printed in the USA
LVHW111037181118
597555LV00006B/239/P

9 781545 644331